THE BUTTER BUSINESS

OR
How a Butter Tub Cover is Related to:

the Mill,
the Cooper,
and the Farmer.

By Caroline Meek
With Modern Illustrations by Joshua de Marrais and Caroline Meek

Hanford Mills Museum, East Meredith, NY 13757

ISBN 1-879444-03-8

For information, contact:

Hanford Mills Museum
P.O. Box 99
East Meredith, NY 13757
(607) 278-5744

CONTENTS

Introduction to Hanford Mills Museum & its Archives

Hanford Mills Museum, situated in the village of East Meredith, New York, owns the original site and buildings of a combination sawmill, gristmill and woodworking shop that operated nearly 150 years. It has some 16 historic buildings including mill related structures, barns and the Hanford family farmstead.

East Meredith started as an unorganized settlement known as Briar Street in the 1820s or 1830s. A local farmer, Jonathan Briggs, built the first section of the mill - an up-and-down sawmill, in 1846. The mill went through a series of owners until a 25 year old carpenter's apprentice and farmer by the name of David Josiah Hanford bought the mill in 1860. After the war, Briar Street received a post office in 1868 and officially became the village of East Meredith. Levi Hanford, D.J. Hanford's uncle, became the first postmaster.

In 1868, D.J. Hanford began the expansion of his mill, completing a gristmill addition in 1869. D.J. continued expanding in 1876, adding the first section of his woodworking shop to make butter tub covers. From this point, the business grew rapidly as the Hanford family added machinery to cut wooden shingles, make crates and boxes, turn broom and tool handles, produce electricity, and improved the existing sawmill and gristmill. D.J. and his sons also added more reliable power sources including a steam engine and later gasoline engines. D.J. Hanford died in 1899. Two of his sons, Horace and Will, took over the mill business, while the third son, John, took over D.J.'s dairy farm. Horace Hanford was the manager of the mill and continued with his father's innovations. The Hanford business became a retail center as well as a mill, selling farm equipment, coal, hardware, and other related items.

Will Hanford died in 1929, leaving Horace the sole owner of the mill. John Hanford died in 1938. His wife to ran their farm until she sold it in 1945. Horace continued in the mill and retail business, although the retail side was the most profitable. Finally, in 1945 at the age of 75, Horace sold the mill to three of his employees, the Pizza brothers - Mike, Frank and Joe. The Pizza's continued the business, though over time they focused more and more on the retail sales.

The last brother, Joe Pizza died, and in 1967, the business was sold to a man named Kenneth Kelso. Mr. Kelso preserved the mill as a private museum and worked to return various parcels of land that had been sold from the original site. In 1973, he sold the site to a non-profit corporation and the Hanford Mills Museum at East Meredith, Inc. came into being, receiving an educational charter.

Today, Hanford Mills Museum maintains the mill as an operating sawmill, gristmill and woodworking shop allowing visitors to get a glimpse of how work was done in the past. The John Hanford Farmstead allows people to see facets of a 1920s dairy farm. Exhibits, films, school programs and special events help visitors explore the relationships between the mill and East Meredith, as well as between the people of the area and the various jobs they were involved in. The Butter Business was a large part of what made East Meredith and the Hanford Mill, and various sides of this business can be seen at Hanford Mills Museum.

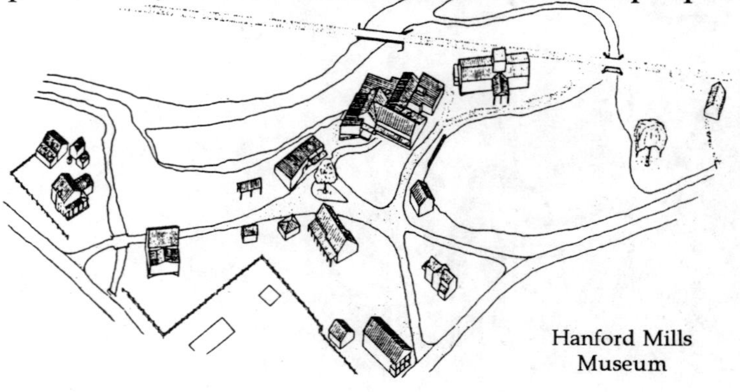

Hanford Mills Museum

Hanford Mills museum is lucky to have a nearly complete collection of the Hanford business records. There are documents from the mill's beginning in 1846 until its closing in 1967. The archives also include documents about local East Meredith area history, historic photographs, common local periodicals and newspapers, documents from other mill and lumber businesses, and a large library of historic and modern books about mills and other related subjects. These archives are open for research, Monday through Friday, 9 a.m. to 5 p.m., year-round by appointment. During the museum's open season from May 1 to October 31, appointments may also be made for Saturday or Sunday, 9 a.m. to 5 p.m.

There are three sections of the Hanford Mills Museum Archive collections that were especially helpful in producing this book. They are listed below with a little background information:

Hanford Mills Business Records

D.J. Hanford kept some records of his early work at the mill. Once his son, Horace, joined the business around 1889, the number of records explodes. The early years of butter tub cover manufacturing are shadowed, but the later years are well documented. The archives contain order books recording covers made; letters sent by coopers with orders, inquiries, complaints and excuses; copies of the letters the Hanfords sent in return; and receipts for machinery the Hanfords bought. With these documents, we can put together a good view of what the Hanfords' butter tub cover business was like.

Trade Catalogs

As a retailer, the Hanford business needed a multitude of catalogs from various sources from which they and their customers could make their choices. Today, a good representational sample of these catalogs have survived. These catalogs cover items such as wagons, engines, mill supplies, farm machinery, radios, office supplies, etc. Catalogs related to the butter business include the Vermont Farm Machine Co., J. Maynes, the De Laval Separator Co., the Sharples Separator Co., Davis & Rankin, Crystal Creamery Co., and many more.

Elizabeth A. Brownell Hanford Diaries

The best source of information about the village of East Meredith, besides the local correspondent column of the Delaware Dairyman newspaper, are the diaries of Elizabeth Hanford. Elizabeth Brownell was born in 1827 and married Levi Hanford, D.J. Hanford's uncle, in 1849. Together they had one son, Charles. The Hanford Mills Museum Archives has in its collection a series of diaries written by Elizabeth covering the years 1863-64, 1866, 1878-86, 1888, and 1890-1902. While the entries are small due to the format of the diary books, Elizabeth managed to pack a lot of useful information about her family, their farm and their community into her entries. She was most active in her family's dairy business in the years from 1863 through 1885. In 1885, Elizabeth was 57 years old and it appears that she gave most of her butter making work to her daughter-in-law, Caroline Flower Hanford. After 1885, Elizabeth occasionally mentions butter churned or sold, but not in detail as she did before. Elizabeth's husband died in 1900 at the age of 77. Her 1902 diary was the last one she kept. She died in 1909 at the age of 81.

ACKNOWLEDGMENTS & EXPLANATIONS

This book is an introduction to the butter business in the East Meredith, New York area. Please read it as an overview, not as an in-depth study. It grew out of a need to help tour guides at Hanford Mills Museum. They demonstrated how Hanford machinery could make a butter tub cover, but most had little idea of what the cover was used for, who used it or where it went after it left the mill. As I worked on The Butter Business, I wanted to include a lot of historic illustrations and information to make their job of studying for tours more enjoyable. When I decided to publish this book for you with the help of the Early American Industries Association and Hanford Mills Museum, I hoped you would find it enjoyable, too. Please read it if you want an introduction to butter, or you wish to learn about the connections between a group of historic businesses, or if you enjoy looking at historic illustrations from the 1800s. I hope that teachers can use it to educated their classes about communities, resources, historic industries, interrelationships, agriculture and the dairy industry. Also, please expect to find mistakes, misinformation and missed chances.

With that said, I would like to thank the following organizations and people. All of them were important to my work, and I couldn't have finished this book without them:

~ The Grants-In-Aid Committee of the Early American Industries Association of Wilmington, Delaware for providing funding for the publication of this book.

~ The Chicago Historical Society Library in Chicago, Illinois for helping me find East Meredith exhibitors at the Chicago World's Exposition. It was thrilling to hear familiar names read to me across time and space.

~ Mary Ann Welsch of Old Sturbridge Village in Sturbridge, Massachusetts for her interesting and informative presentation on historic cheese making at the "Union of Spirits Conference for Interpreters - Food For Thought II" 1994 at Mystic, Connecticut.

~ The Association for Living Historical Farms and Agricultural Museums 1994 Troy, New York conference for letting me do a presentation on butter tub covers which helped me refine my ideas for this book.

~ Gwen Miner of the Farmers' Museum of Cooperstown, New York for her support and all her dairy information help.

~ Susan R. Perkins of the Herkimer County Historical Society in Herkimer, New York for her information on New York's dairy exhibit at the Columbian World's Exposition.

~ Joshua de Marrais for his graphic skills, who helped provide illustrations for those topics where a historic drawing just wouldn't work.

~ And Hanford Mills Museum and its staff for providing me the support and archives to do his book, and for putting up with the time it took me to finish it.

What is a Butter Tub Cover?

When visitors come to Hanford Mills Museum, they see many of the water powered machines producing products that were made in the late 1800s and early 1900s. One of these products is the Butter Tub Cover. To many people, a Butter Tub Cover may look like a frisbee or a stool seat or even a toilet cover. Why should anyone be interested in this round piece of wood?

The Hanfords were making these covers as recently as 80 years ago. Many of your grandparents or your parents or even you remember seeing them or using them. But, in today's society, they are forgotten. Eighty years ago, they were very inportant for dairy producing areas, especially those places like the East Meredith area where butter was made and shipped. The Butter Tub Cover was important to more people than just the mill worker who cut it.

It often happens that when you understand more about a thing, you also find it more interesting, so this book was put together to help people understand why they might want to know about these round pieces of wood, and who might have used them. The book is divided into four chapters. Chapter 1, The Mill tells about how the covers where made. Chapter 2, The Cooper tells who bought the covers and what they used them on. Chapter 3, The Farmer tells about what was packed under the covers and how it was used. And finally, Chapter 4, The End tells about why few people recognize covers anymore.

So, you are probably still asking yourself - What is a Butter Tub Cover?

A Butter Tub Cover is a round wooden lid made to cover a Butter Tub which was used to store and ship butter.

If you want to find out more, please read on ...

What is a Butter Tub and a Butter Firkin?

It is difficult to say exactly what either a butter tub or firkin looked like. Sizes and names seem to vary from place to place - even within the area around East Meredith, New York. To this date, we have not found any authentic butter containers. According to many older visitors, most tubs and firkins were broken up after use and burned since they were not reused for butter again. Here is what we have found:

What We Know About Butter Tubs:

1. Butter Tubs were made by coopers. They looked like a wooden **pail or bucket**.

2. The Hanfords advertised maple, beech and birch tub covers, but sold mostly **ash, maple and oak.**

3. We have no information on what butter tubs were made of in this area, but according to national advertising, tubs were made of **ash or spruce**.

4. While butter tub sizes varied, both D.J. Hanford and his aunt, Elizabeth Hanford, packed their butter in tubs that held an average of **50 pounds.**

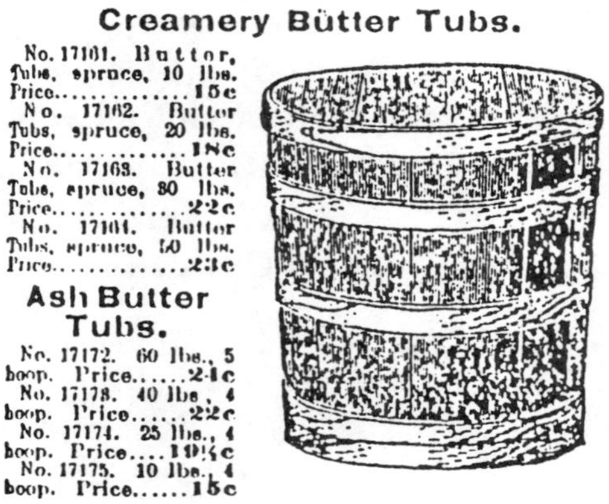

Sears, Roebuck & Co.
1897

Ash Butter Tubs.

A.H. Reid
Date Unknown

What We Know About Butter Firkins?

1. They were also made by coopers, but they looked like **barrels**. Firkins were literally cut in half to make **half-firkins**.
2. Firkins were probably made of materials similar to butter tubs.
3. Elizabeth Hanford packed butter in firkins that held an average **100 pounds**.

Firkin and
Half-Firkin

MYSTERY:

Lewis Quick of Meridale, N.Y., who was about 80 when he died in 1991, had a different definition of "firkin." He described a firkin as a container made of wrapped wood. In the old advertising, there is a container with that description - a **Butter Box** made of wrapped wood much like cheese boxes. According to old catalogs, different sizes were made to hold between 1/2 to 10 pounds of butter. The Hanfords made and sold something they called a butter box, but there is no information whether it was round or square.

~ How wide spread was Lewis Quick's definition of "firkin"?
~ What sort of butter box did the Hanfords make and was it ever called a firkin?

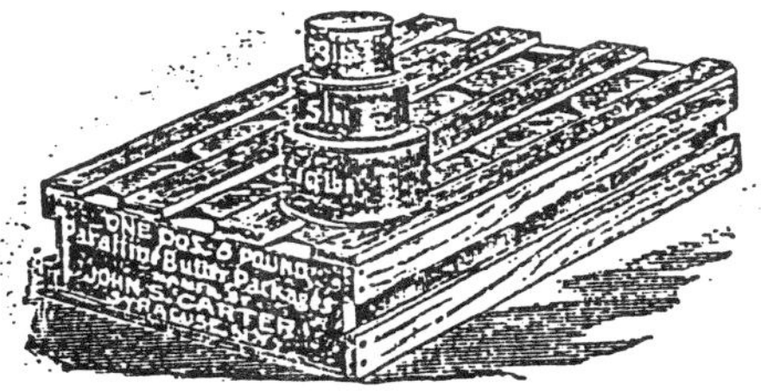

Butter Boxes
John S. Carter
Date Unknown

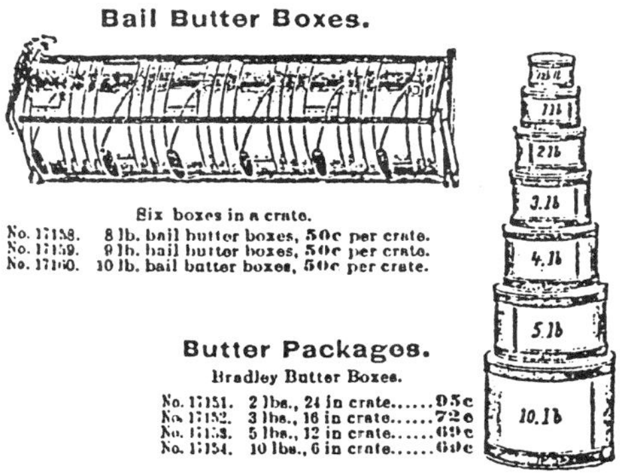

Sears, Roebuck & Co.
1897

Chapter 1

The Mill:

IMPORTANT DATES - Hanford Tub Cover Timeline

1876

Before September

Some time before September, D.J. Hanford bought the mill's first Tub Cover Cutter. It was a machine patented by W. Mickel of Oneonta (see the Machines information). There are two pieces of evidence that tell us D.J. bought the machine in 1876.

EVIDENCE:

1. D.J. Hanford's records show that the first Butter Tub Cover order was made in September of 1876.
2. D.J. Hanford used a small pocket diary to record mill orders, then he transferred the information to a permanent Daybook. The 1876 Diary has a drawing on the back inside cover of a shaft that may represent the Mickel Tub Cover Cutter.

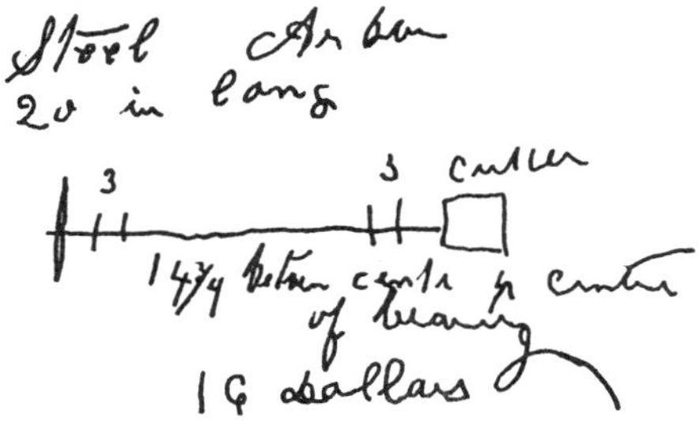

D.J. Hanford Diary
1876

Is this the Mickel Tub Cover Cutter?

When the Tub Cover Cutter arrived, the mill had only two sections (the sawmill and the gristmill areas). The Cutter was probably set up in the sawmill.

EVIDENCE:

1. In the Spring of 1991, while cleaning the top of a shavings bin built above the sawmill, museum staff found the corners that the cutter makes when it cuts a butter tub cover. From the museum's experience with the Mickel machine, we know it can throw corners that high when its operating.

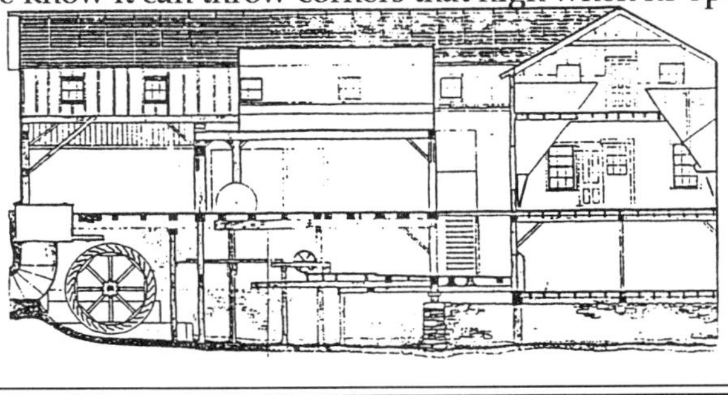

September 9

D.J. Hanford entered the first Butter Tub Cover order in his Daybook - an order for 435 covers. D.J. only cut a little more than 1000 covers his first year.

1881

Before October 21

On October 21, D.J. Hanford and his cousin, Charlie Hanford, went to Oneonta to pick up the mill's first Steam Engine. They installed it in the basement of the basement of the Tub Cover Room. This addition to the mill was probably built just before the engine was added. Shafting was installed that could power the Tub Cover Cutter either with water power from the turbine or with steam power from the new engine. D.J. cut about 13,500 covers in 1881.

The Mill before the tub cover room, c. 1874.

The Mill after the addition of the tub cover room, c. 1894.

1888

August 7

D.J. Hanford ordered his second Butter Tub Cover Cutter (it was called a Square Edge Turner by the company - see the Machines information). D.J. cut a little over 37,000 covers in 1888.

1916

June 5

D.J. Hanford's son, Horace, recorded the Last Butter Tub Cover Order in the daybook. It was the only order recorded for 1916. George W. Sides ordered 150 covers. In 1915, the year before, Horace cut only 1000 covers.

1965

August 7

The last owners of the mill, the Pizza family, have a Going-Out-of-Business Auction, and sell many of their machines. The Butter Tub Cover Machinery is bought by Kenneth Kelso, who bought the mill in 1967 to run as a museum.

1987

March

The Mickel Tub Cover Cutter is restored and begins operation in the museum.

1991

November

The Matcher (see The Machines) is rediscovered at the museum. It is restored to its original location and begins operation.

THE MACHINES - The Mickel Barrel Head Machine

Other Names:	Patented Head Cutter, Tub Bottom Lathe or Square Edge Turner
Patented By:	William Mickel of Oneonta, New York
Patent Date:	May 31, 1870
Patent Number:	103,762
Manufactured By:	Ford & Howe of Oneonta, New York(?) Reorganized in 1871 as the Oneonta Agricultural & Machine Works

In 1873, the Oneonta Agricultural & Machine Works listed "Mickel's Patented Head Cutter" as one of its products, but the company closed down soon after 1873 (according to Oneonta: The Development of a Railroad Town by Eugene D. Milener).

MYSTERY:

D.J. Hanford started butter tub cover production in 1876, nearly two to three years after the Oneonta Agricultural & Machine Works went out of business.
Who did D.J. Hanford buy his Head Cutter from?

- Was Ford & Howe or the Oneonta Agricultural & Machine Works still selling equipment in 1876?
OR
- Did D.J. buy the machine in 1873 or before, when the company was known to be in operation, and then not use the machine until 1876?
OR
- Was someone else manufacturing machines similar to Mickel's?
OR
- Was Mickel making his own machine?

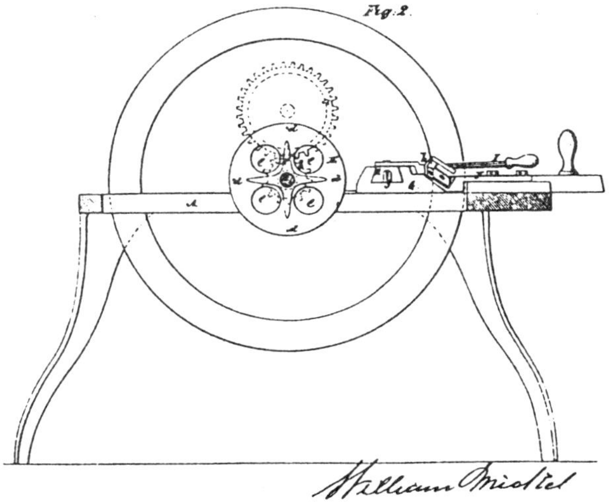

There are some changes from Mickel's 1870 head cutter and the machine at Hanford Mills Museum, but the cutter heads and the clamps are nearly the same. Perhaps William Mickel or the manufacturers made improvements to the machine that were either not patented or the museum has not found the records at this time.

W. Mickel,

Manf. Barrel Heads.

No. 103,762. Patented May 31. 1870.

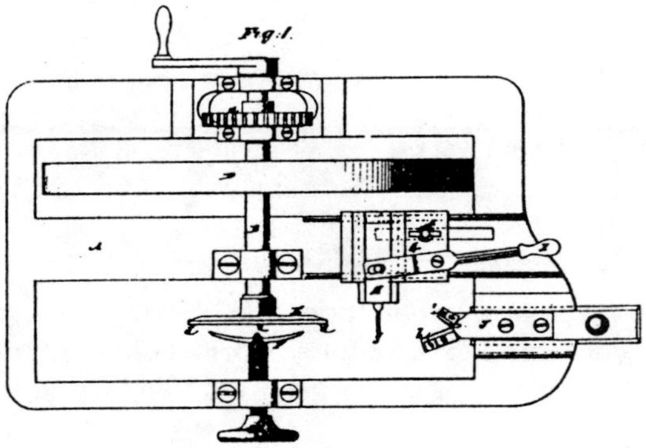

Trivia - Trivia - Trivia - Trivia - Trivia - Trivia - Trivia - Trivia - Trivia

Timed tests done by Hanford Mills Museum staff indicate that the Mickel machine could turn about 1,000 to 1,200 covers in one 10 hour day. At that rate, the Mickel machine would produce a stack of butter tub covers that was 41 to 50 feet tall in one day! That is when both the machine and operator are at their best. It is not known whether the mill ever made this many covers in a day.

The Mickel Barrel Head Machine requires many steps to cut a barrel head. The wood blank must be clamped in the machine and then it must be turned on. Next one knife is slid in to cut the circle and then a separate knife is used to cut the bevel. The machine must then be turned off and the cover removed.

THE MACHINES - The Matcher

Other Name: Tongue and Groove Machine

Very little information is known about this machine. It was "rediscovered" in the Hanford Mills Museum collection in November of 1991. It had been stored in the basement of the museum's Feed Mill with other machinery. Many of the other machines are donations, and are not original Hanford machines.

Why Do We Think This is a Hanford Machine?

1. In the Tub Cover Room of the mill there are **Bolt Holes** in the floor that match the bolt holes in the matcher. Since these holes are asymmetrical, the staff was able to place the machine exactly as it was originally. There are also two **Wear Marks** in the floor where the operator stood to run the machine.

2. A missing shaft from the museum's incomplete Hanford Trevor Square Edge Turner (see next section) was found on the matcher.

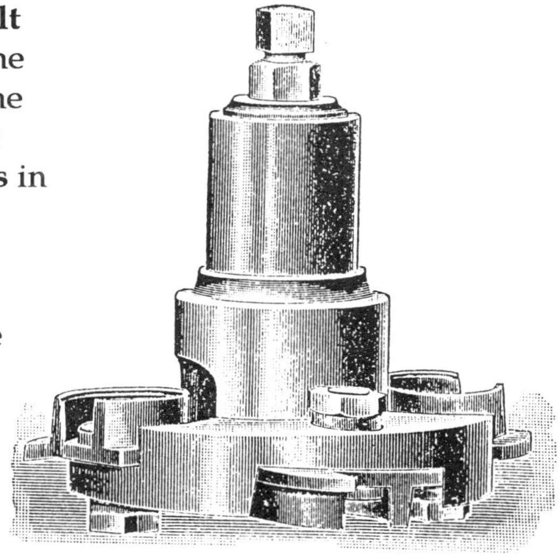

4 Bit Groove Head

What Do We Know About the Matcher?

1. This machine once had a jack shaft at the bottom of its frame to run the upper shaft. It is no longer in existence and was probably removed by the Hanfords since the **Belt Holes** in the floor indicate that the machine was set up to be directly powered from below.

2. The frame of the matcher was once painted **Green** (The Mickel Head Cutter may also have been green).

3. The matcher is fitted with **Shimer Matcher Heads** made by the Samuel J. Shimer and Sons Company of Milton, PA. These are not the original tongue and grove cutter heads since the last patent date is 1885, but Shimer was making similar heads before that date.

4 Bit Tongue Head.

Complete Shimer heads are illustrated on the previous page, but this quote, taken from a 1929 Samuel J. Shimer and Sons catalog, explains how the cutter heads work:

The Shimer cutter Heads since their introduction in 1873 have grown into general favor by reason of their simplicity, convenience and economy.

As applied to the manufacture of flooring the Circular Bits are arranged alternately on opposite sides of the flange having seats therein and Bits on the seats secured by bolts and nuts. In operation these Bits are divided and work in upper and lower series cutting the chip alternately with suitable clearance from a divided line of cut.

First one Bit makes a cut, then the next Bit of the other series follows. The entire cut is therefore, not made at one carve. Both slightly shear - dividing the chip along the center line of cut.

Permit us to illustrate by diagram, Fig. 3, showing groove Bit D in the position it occupies when making the cut. The Bit C which follows to complete the work, is given in outline. This shows how the chip is divided, and from it you will understand that the several Bits that thus complete the work are arranged in upper and lower series, upon a Head, that has a series of seats to which these Bits are secured.

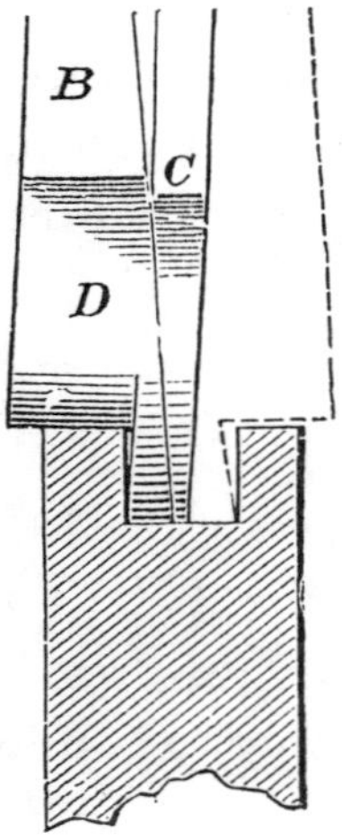

Fig. 3.

The Bits are inclined alternately to and from a dividing line central to this arrangement of seats. **The object of the inclining seats, is to give to the Bits thus arranged a side clearance, and they take a position the same as a saw tooth when set.** <u>You will observe, therefore, that the Bits when at work strike the lumber with their edges only.</u>

Having a clearance thus carefully and permanently provided in connection with the divided ship, the Bits turn out smoother work on hard, cross-grained and knotty lumber.

THE MACHINES - The Trevor Patent Heading Turner

Other Names:	Square Edge Turner or Centennial Heading Turner
Patent Date:	July 27, 1875
Manufactured By:	Trevor & Company of Lockport, New York (Also known as the Trevor Manufacturing Company)
Purchase Price:	$212.00
Purchase Date:	August 7, 1888

The Trevor is more mechanized than the Mickel machine. The operator only had to put the wood between the clamps and step down on a lever. A universal joint closed the clamps and tightened the drive belt within the machine. The arbor turned, allowing a concave saw to cut both the circle and bevel in one and a half turns (the machine could be set up to cut square edges - i.e. cut just the circle and not the bevel). As the head was cut, a pin traveled a circle on an attached gear. When the head was cut, the pin hit a release lever, the tub cover rolled out and the machine was ready to start all over again.

All goods at buyers risk after shipment.

Lockport, N.Y. Aug 7 1888

ESTABLISHED 1860.

TREVOR & CO.

IMPROVED MACHINERY FOR SHINGLES, HEADING, STAVES, HANDLES & WOOD PULP.

WORKS 166-168-170-172 MARKET ST.

TERMS CASH. SUBJECT TO SIGHT DRAFT WITH EXCHANGE.

Sold to D. J. Hanford East Meredith N.Y.

1	Square Edge Turner	235 00	
1	Counter Shaft for do 14' long Extra hanger	30 00	
		265 00	
	20%	53 00	212 00

Please remit by draft on New York or Post Office order. If local check is sent add 25cts. for expense of collecting the check.

Shipped by R.R.

MYSTERY:

The original Trevor invoice for the Hanford machine calls it a "Square Edge Turner." This suggests that their machine did not cut bevels, but what kind of square edge turner was it?

According to advertising, there were a number of ways to attach the cutting mechanism to the machine, depending on the exact name of the heading turner.

Which way was the Hanford Trevor machine set up?

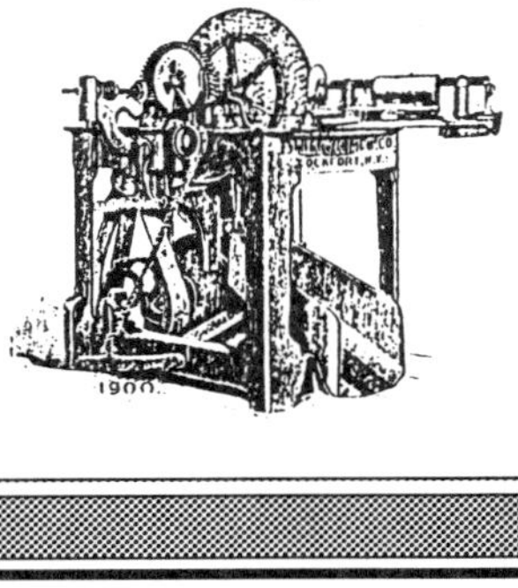

Trevor Manufacturing Co.
1910 Catalog
1900 Model

THE MACHINES - Con't.

The Hanford's Trevor Heading Turning is mentioned twice in the local paper - The Delaware County Dairyman. This paper covered all of Delaware County and included a section for each village.

Delaware Dairyman

JOS. EVELAND, Publisher.

Friday - - September 21, 1888

D.J. Hanford has put a new head cutter in his tub cover manufactory. He can now cut seven covers a minute on an average.

Friday - - December 28, 1888

Below the store is D.J. Hanford's mill, complete in all departments and fitted with the latest improved machinery including a planer that will plane both sides of a board at once and also deliver its shavings at another part of the mill where they are carried off by the water from the wheels, and a machine for cutting tub covers with a capacity of 3500 covers per day. Mr. Hanford does an extensive business in tub covers.

The math does not quite add up between September and December, but perhaps D.J. had a better idea of exactly what his new heading machine could produce after using for a time. Consider the following math:

September: ~ 60 minutes X 7 COVERS PER MINUTE = 420 covers per hour
~ 420 covers per hour X 10 hour day = **4200** COVERS PER DAY

December: ~ **3500** COVERS PER DAY / 10 hour day = 350 covers per hour
~ 350 covers per hour / 60 minutes = **5.83** COVERS PER MINUTE

Trivia - Trivia - Trivia

At 3500 covers per day, the Trevor Heading machine could make a stack of tub covers that stood 145 feet and 10 inches tall!

Woodworker magazine, May, 1893

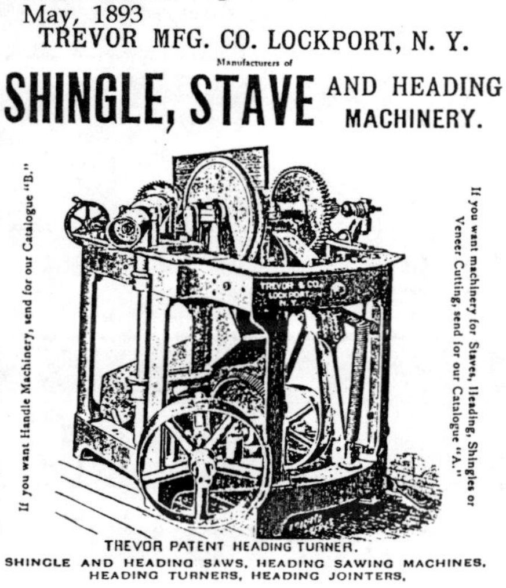

Trevor Manufacturing Company, 1910 Catalog
1907 Model

MYSTERY:

For some reason, after the Hanfords stopped making tub covers in 1916, the Trevor machine was partially disassembled and then stored away.

- To this date some pieces are still missing. Why?

Heading Specifications - Woods

One of the first ads that mention the wood D.J. Hanford use to manufacture tub covers is from the East Meredith Gazette newspaper published on January 1, 1880.

D. J. HANFORD
East Meredith, N. Y.,
GENERAL DEALER IN
Flour, Feed and Lumber.

MANUFACTURERS OF
White Maple Tub Covers.
I AM PREPARED
To furnish Tub Covers in large or small quantities, at prices which defy competition.

In this quote from a letter the Hanfords sent to a prospective customer in March of 1894, they give a more complete description of what their covers were made of:

> *They are made from Maple, Beech and Birch principally and practically clear. Of course not absolutely white, but good sound timber only is used.*

The woods mentioned in the order books and customer orders do not match what the Hanfords advertised. Not every order mentions materials, but of those that do, these examples show how these materials were divided among different wood types:

Orders from 1893 to 1895

50%	Ash
50%	Maple

Orders from 1901 to 1902

40%	Ash
40%	Maple
17%	Oak
3%	Beech

D. J. Hanford & Sons,
DEALERS IN
Flour, Feed & Lumber,
Agricultural Implements, &c.
Manufacturers of
Firkin Heads, Tub Covers
AND HANDLES.
EAST MEREDITH, N. Y.

Hanford Business Card
About 1901 to 1905

Sometimes Ash is specified as White Ash, and at least one Oak order is specified as Red Oak. Only one other wood is mentioned, and only in one order - Elm. Birch, to this date, has never been found mentioned in an order, though the Hanfords wrote they used it.

As the Hanfords stated, they used "practically clear" wood for their covers, but that does not mean it was prime lumber. Butter Tub Cover were meant to be used and thrown away. The Hanfords probably used clear boards from waste or lower grade wood. Using wide boards was expensive, and so some sort of joint was necessary.

Each of the woods that the Hanfords used to make tub covers, or said they used, have specific properties (as listed in Important Forest Trees of the Eastern United States, published by the United State Forest Service). These woods are listed below in alphabetical order. Notice that all of these wood types have similar characteristics:

ASH:

Properties:
Heavy, hard, strong, stiff, has good shock resistance and excellent bending qualities.

Uses:
Handles, furniture, **dry cooperage** and sports equipment.

Compound, opposite leaves; 7-9 leaflets; light brown bark with a diamond pattern of ridges; young bark is smoother and light orange-brown.

BEECH:

Properties:
Strong, heavy, shock resistant and readily bent when steamed.

Uses:
Lumber, veneer, ties, pulpwood, **dry cooperage** and fuel - **is especially good for containers since it does not transmit taste or color to container contents.**

Simple, pinnate, alternate leaves; smooth, light gray bark with find black spots.

BIRCH:

Properties:
Heavy, hard, strong, stiff and very high shock resistance.

Uses:
Lumber, veneer, distilled products, crossties, pulpwood and **dry cooperage.**

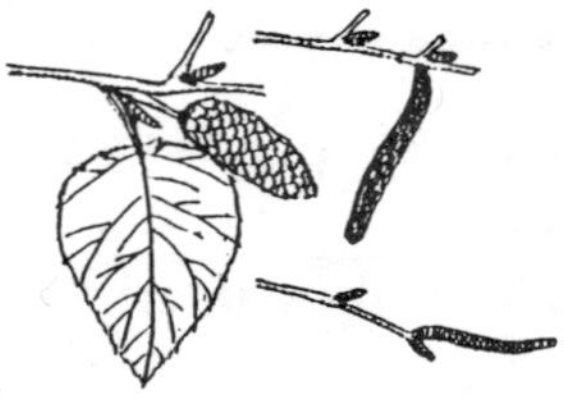

Simple, pinnate, alternate, oval leaves; white bark that peals in horizontal strips.

ELM:

Properties:

Moderate in heaviness, hardness and stiffness, has good shock resistance but is moderately weak and has excellent bending qualities.

Uses:

Containers, furniture, **dairy** and poultry supplies and **dry cooperage and kegs.**

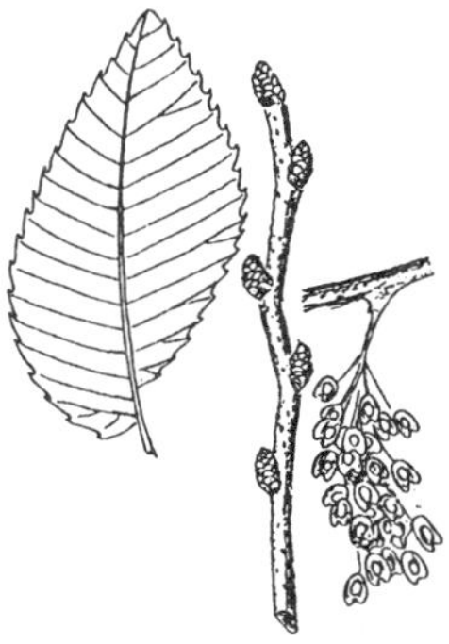

Simple, pinnate, alternate leaves with uneven bases; sliced or broken bark has alternating light and dark layers.

MAPLE:

Properties:

Heavy, strong, stiff and high resistance to shock.

Uses:

Lumber, **boxes and crates**, distilled products, veneer, crossties and pulpwood.

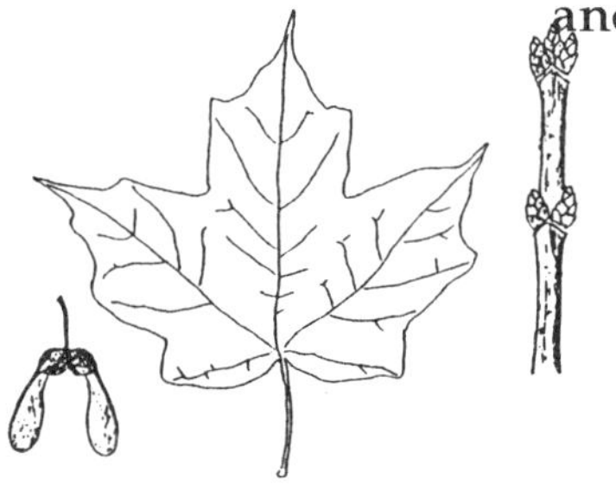

Simple, palmate, 5-lobed, opposite leaves; vertically furrowed bark with flat-top ridge plates; young bark is dull gray.

OAK:

Properties:

Heavy, hard, stiff and high resistance to shock.

Uses:

Flooring, **dry cooperage (Red Oak - due to being extremely porous) and wet cooperage (White Oak)**, furniture, boxes, timbers, handles, coffins and boats.

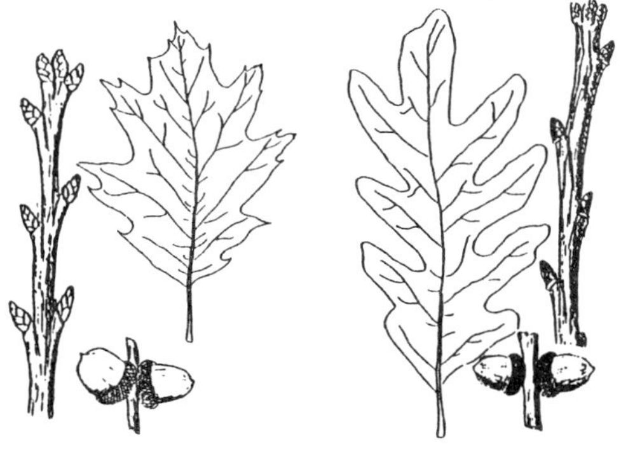

Simple, pinnate, alternate leaves with 7-11 pointed lobes and bristled tips (Red) and flat, smooth-topped ridges on older bark; or 5-9 rounded lobes and flat, scaly, older bark (White).

HEADING SPECIFICATIONS - Joints

When Hanford customers ordered their tub covers they had a choice of how they wanted the cover pieces joined together. Their choices were matched, pegged, jointed or flagged:

MATCHED:

Almost all Hanford tub covers that exist today have **matched joints (also known as a tongue and grooved joint)**. Heading blanks were matched on the matcher (see the Machine information) before they were cut into the circular shape.

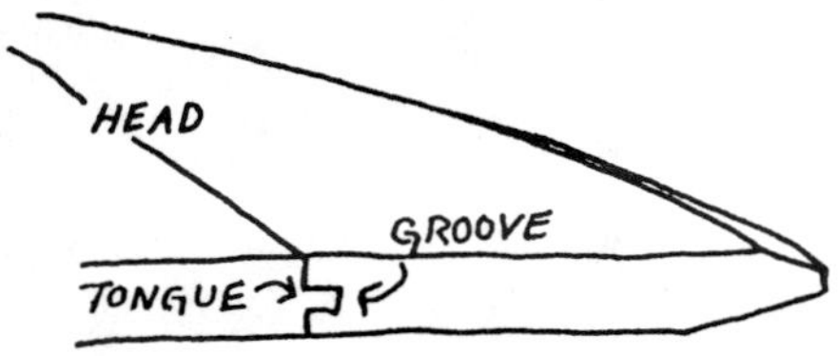

Many people ask if heading material was **glued** together. The Hanfords did not glue, since the tongue and groove fit snugly. When the head got wet, the joint would swell even tighter. The Hanfords received one letter that actually mentions glue (the emphasis is Mr. Sloan's):

> *Walton Apr 4th 1899*
>
> *D.J. Hanford & Sons*
> *Dear Sirs.*
>
> *Please Send me 150 Tub Covers 16 1/4 and 50 16 6/8 & 50 15 3/4. I want them made full 1/2 an inch thick <u>and no Glue</u> in the Grooves. Just the kind I got from you formerly. Please send them as soon as possible and Oblige*
>
> *Yours Truly*
> *R.H. Sloan*

PEGGED:

The museum has one example of a pegged or doweled head, but there are no recorded orders that specify pegged heading. Covers were cut with straight joints, holes were drilled, and pegs placed in the holes to hold the cover together.

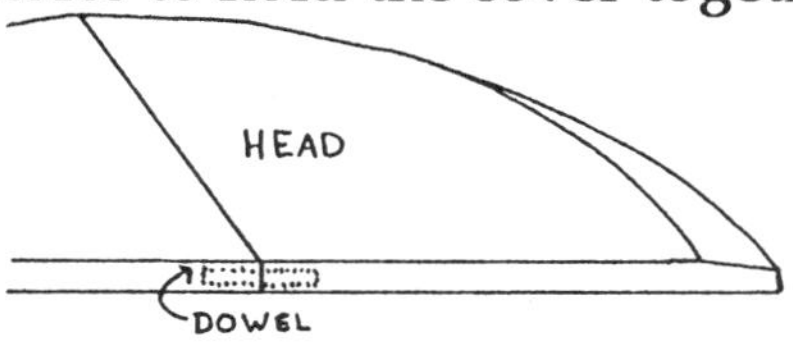

While none of the Hanford orders specify pegged heads, the museum does have a few boxes of quarter inch diameter, two inch long dowels or pegs. There are orders for jointed heads with pegs separate, and the cooper put the heads together himself:

Warnerville
April 5 99

Mr Hanford
Dear Sir
have you ash headin that is seasond as I am in nead of some for tub heads that is 5/8 thick and 14 inches long if so send me anuf for 100 heads 2 peaces for a head as I dowel them to gather Send them as soon as posable and i will send you check for the same
Ship to Cobelskill
Yours Truly
L. Terrell

Bovina Center
Oct. 9th 1901

Mr Hanford
Dear Sir can you ship me 300 covers ship as soon as you can and 1 bbl of ash heads and 4 quarts of dowel pins put in bbl of heads
Resp
Chas Palmer

"bbl" is an abbreviation for barrel.

The museum also owns a tool that was used to put peg holes in heading. Since one pegged cover exists in the collection, the Hanfords may have pegged some of their own heads. The doweling tool is basically two drill bits that can be adjusted for different widths. Both drills are driven by one hand crank. A heading blank was placed on the platform in line with the drills. Once the holes were drilled, pegs were fitted in and the blank was put together.

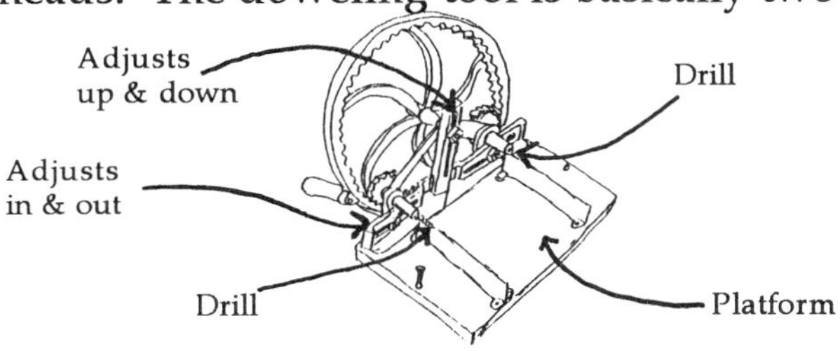

JOINTED:

Sometimes there were orders for heading that was neither matched or pegged, just straight **jointed**. The two orders above would have been jointed.

FLAGGING:

At times, especially for containers that were going to hold liquids, the Hanfords were asked to **flag** their heads. This meant taking a rush or flag (such as grows in wetlands) and placing it between the joints. When wet, flagging would swell and seal the head. Flagging expands faster than wood. Here are some related quotes from letters to the Hanfords:

May 10, 1902

Please send me 200 maple covers size 16 1/4 also 200 ash heads. I would like the heads flagged.

September 16, 1903

I will want 200 Pork Barl heads with the next order They must be Flaged in Joints if you have not Flag write and I will send you some

HEADING SPECIFICATIONS - Types and Sizes

Heading was ordered using one or more of the following names. Each of these terms comes from the Hanford record books and order letters, and these same archive records suggest the definitions given below. Other regions of the state and country may have had other definitions for these same words.

Bottom	The bottom of a butter tub (also called a "head").
Cover	The top of a butter tub (also called a "lid").
Dry Head	The top or bottom for a barrel that held dry materials, and thus, did not have to be watertight construction.
Firkin Head	The top or bottom of a butter firkin.
Flour Barrel Head	The top or bottom for a flour barrel which did not have to be watertight.
Head	The bottom of a butter tub (also called a "bottom") OR in general, the top or bottom of a container from a cooperage.
Lid	The top of a butter tub (also called a "cover").
Pork Barrel Head	The top or bottom for a barrel built to hold pork in brine, this heading was flagged and had to be watertight.
Square-Edge Cover	A butter tub cover without a beveled edge (also called a "square head").
Square Head	A butter tub cover without a beveled edge (also called a "square-edge cover").
Uncut Head	Either a butter tub cover without a beveled edge or a square blank, matched or jointed, but not cut in a circle (records do not give enough information to be sure).
Welsh Head	???????????? - this is a term used only in early orders, and there is no information to help define the term.

The Hanfords could make many sizes of heading, and orders were taken for anything between 9 $^{5}/_{8}$ inches to 19 $^{1}/_{8}$ inches, with nearly all the increments in between. Firkin and barrel tops and bottoms were the same size, but butter tubs had a larger top. Orders for butter tub covers and heads were often made together.

Only one cooper mentioned the size of his butter tub when he ordered his covers and heads. He wrote that he wanted:

14 inch covers & 10 $^{5}/_{8}$ inch heads for his **30 pound butter tubs**

This cooper's order suggests that the following sizes, commonly appearing in orders were for **50 pound butter tubs**:

Covers	16 $^{1}/_{4}$	16 $^{1}/_{4}$	16 $^{3}/_{8}$	16 $^{1}/_{2}$	16 $^{3}/_{4}$
Heads	12 $^{1}/_{2}$	12 $^{5}/_{8}$	12 $^{3}/_{4}$	12 $^{5}/_{8}$	12 $^{5}/_{8}$

These orders also suggest that every cooper's 50 pound butter tub was not exactly the same height.

A fifty pound butter tub with the following size heading:

16 $^{1}/_{4}$ inch cover
12 $^{1}/_{2}$ inch head

was probably about **12 inches high**. Tubs with other size heading would likely be about the same height - though a bit higher or lower. A butter tub this size would probably have a silhouette that looked like this:

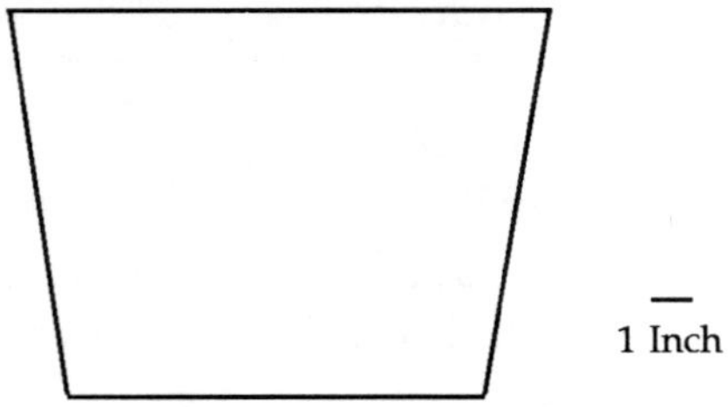

To this date, the museum does not have an original, historic example of a butter tub.

~ Once a butter tub was emptied, it was broken up and thrown away or used as firewood.

~ A butter tub was a **Disposable Container**.

Production Numbers

Hanford Mills Museum has been able to preserve most of the Hanford business records. The following graph is the result of hours of pouring through Hanford order books, picking out only those orders for butter tub covers. To this date, there is one order book missing, and research is incomplete. Unfortunately, there are these gaps in the information, but you can an idea of the number of tub covers the Hanfords were selling each year can still be seen..

Tub Cover Sales Per Year

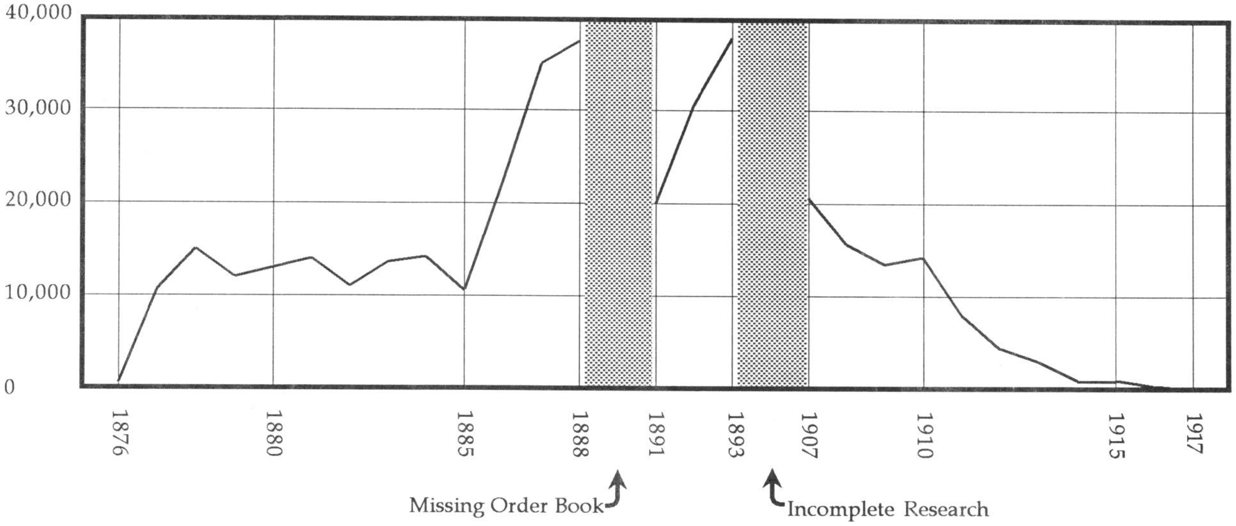

Not only do the Hanford Order Books give information on heading sales through the years, but they also show a pattern of heading sales month to month. You will notice in the graph below, heading sales went up in the spring and late summer. Demand for butter tubs would go up in the spring when cows became available to milk, and in the fall when many people thought about saving the last butter to store for winter use and sale.

Average Tub Cover Sales Per Month

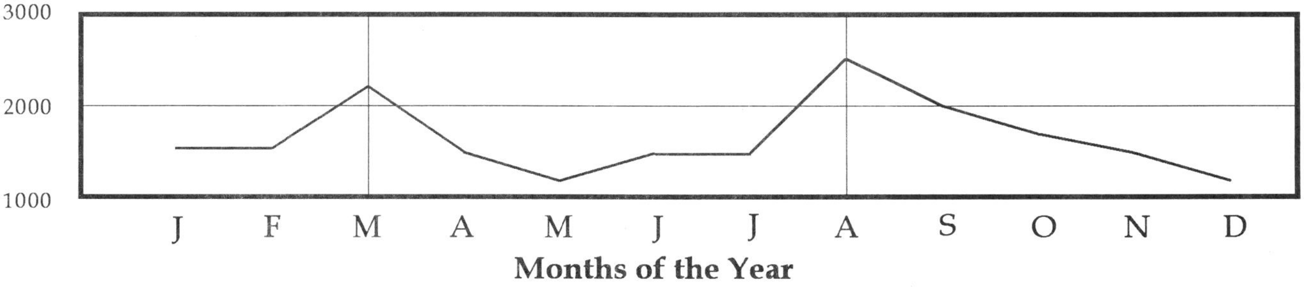

This graph is based on the butter tub cover orders for the years 1877 through 1910 (not including the missing years 1889 to 1890 and 1894 to 1906).

Prices

D.J. Hanford sold his first order of tub covers for a low price. On September 9, 1876, the following order was recorded:

435 covers sold for $130

This tells us that D.J. sold these covers for about 29.8 cents a cover, or nearly **$3 per 100** (which is how his orders were priced - by a 100 covers at a time). This is a low price considering he sold the rest of his orders in 1876 for:

$6 per 100 covers.

By 1877, D.J. Hanford made distinctions in price depending on the materials he used. Price did not depend on what size a cover was, though:

For Maple heading (all sizes)	$5 per 100 covers
For Oak and Ash heading (all sizes)	$6 per 100 covers

By 1900, heading prices varied between **$2.50 and $4.00 per 100 covers**. These prices depended on many factors, such as the materials used, the size of the heading, the degree to which they were finished and even perhaps on how well the customer was known to the Hanfords. At this point, it is too difficult to come up with a copy of the price schedule the Hanfords were using.

DISCOUNTS

In the first year, 1876, a discount of 10% was offered. After that year, discounts ranged from 5% to 15% off. There are many letters similar to this:

August 28, 1895

Bovina, N.Y.
Charles Palmer

Dear Sir:

Answering you inquiry of 26th int. would say, We can furnish you covers for $4.25 per hundred on time and will make a discount of 5% for cash

We would be pleased to furnish you your covers but can not give you more than 300 at present We are repairing and do not expect to be running under two or three weeks.

Please advise us if this will be satisfactory.

Very truly yours
D.J. Hanford & Sons.

SHIPPING

Butter Tub Covers were sent to coopers in three different ways:

1. The cooper picked them up at the mill.

2. The mill delivered them with their wagon.

3. Or, the mill sent them on the railroad.

Shipping was an added cost over the cost of the heading. It depended on destination and method of shipment.

Covers and Heading were packaged differently.

1. **Covers** were delivered in **Bunches of 50.**

2. **Heads** were packed in **Barrels which held about 100 to 110 heads.**
   ~ The Barrel was 10 cents extra.

The following terms also appear in tub cover orders:

1. A Bale appears to be about 100 covers.

2. A Lade appears to be about 1000 covers.

MYSTERY:

The dictionary definition of "lade" is "to load or as with a cargo."

~ Is "lade" used in cover orders as just another word for "a load" or is it an actual cooper or shipping term? 1000 covers would make a hefty wagon load.

Tub Cover Distribution

The Hanfords mostly shipped to what we would call today a local market. They made extensive use of the local railroad lines, and even used them to ship covers to places like New York City and Honesdale, Pennsylvania.

The map below shows Hanford Butter Tub Cover Distribution from 1889 to 1905. Note how shipping destinations are rarely far from a rail-line.

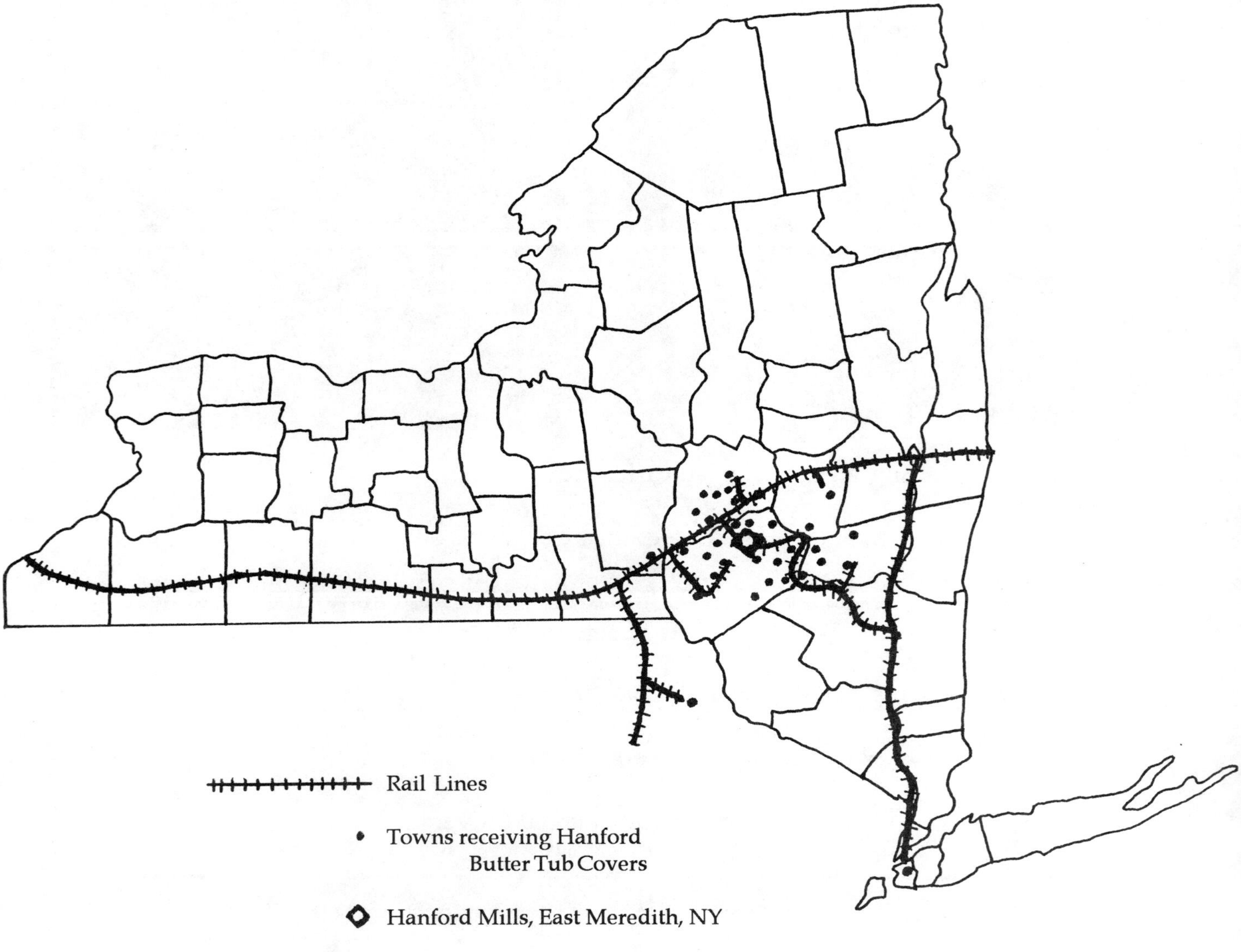

Chapter 2

THE COOPER:

Types of Cooperage

What is a cooper? - A cooper is a person who makes or repairs casks, barrels, etc.
What is cooperage? - Cooperage are the items made by a cooper.

There are three different types of coopers who each made a different type of cooperage:

White

1. These coopers make **buckets, tubs and churns** (usually splayed casks that are often open on one end instead of being an enclosed cylinder with a bulge in the middle).
2. White cooperage is sometimes considered part of Wet Cooperage because **the product is built to hold liquids and is built to last.**
3. This was the first type of cooperage to be affected by industrialization and new materials.

Wet

1. These coopers made **barrels to hold liquids**, like beer, wine, liquor, vinegar, oil, turpentine, etc.
2. Since barrels had to be made to store liquids for long periods of time, **coopers worked with the best woods and built barrels to last many years of use.**
3. Barrels built for alcohol **had to withstand the pressures of fermentation** and repeated use, while other liquid barrels had to withstand repeated use and shipping.
4. Wet cooperage was least affected by industrialization, especially barrels for alcohol, because of the required quality and strength of the product.

Dry

1. These coopers make **rougher types of casks and barrels.** This cooperage can be divided into slack and tight:

    - **Slack barrels** are made for flour, apples, chemicals, nails, packing glass and china, or other dry items.

    - **Tight barrels** are made for holding damp items and liquids for shipping, such as butter, soap, meat, fish, syrup, etc.

2. These barrels are **made of cheap materials** and are meant **to last one trip and then be broken up and discarded.**

3. While Dry cooperage can be considered less exacting than Wet cooperage because the barrels are not meant to last, Dry cooperage is complicated by the variety of barrel types, both slack and tight.

4. Dry cooperage was also affected early on by industrialization. Machines for dry cooperage were easier to make than for wet cooperage. Also new containers were developed for shipping, such as metal drums and cardboard cartons.

Making a Barrel

In his book The Cooper and His Trade, Kenneth Kilby states:

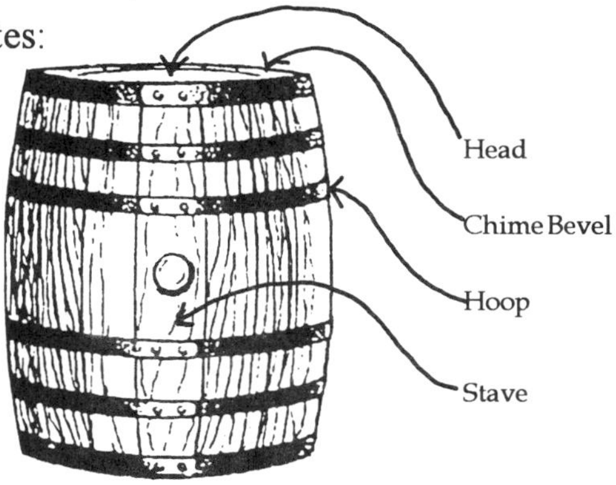

> **Robinson Crusoe was able to make practically anything, but he never made a barrel. There are no amateur barrel-makers. Coopering is a skill acquired through years of sweating, muscle-aching, back-breaking labor.**

The Hanfords never made barrels or butter tubs for this reason. Not only was learning the skills to make barrels a long process, but there are many steps and tools involved, as well. The following descriptions were put together with the help of the book Dictionary of Woodworking Tools, by R.A. Salaman.

Step 1: Shaping the Staves

1. The staves are cut and *listed* or tapered with a **Cooper's Axe.**

A thin, flat T-shaped blade, measuring 10 to 12 inches along cutting edge. It has no poll and is ground on one side only. The handle is offset to prevent the user from hurting his knuckles.

2. The back of the staves are rounded with a **Backing Draw Knife.**

A straight bladed draw knife that is from 10 to 18 inches long. It is similar to a carpenter's draw knife, except the handles are sometimes bent down below the cutting edge to help make deep cuts.

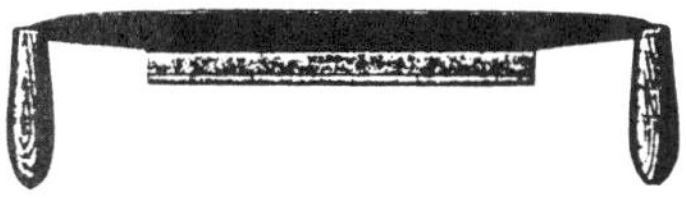

3. The inside of the staves are hollowed out with a **Hollowing Draw Knife.**

A draw knife with the blade bent in a shallow, hollow curve. Usually about 12 inches long and 2 inches wide.

4. The edges of the staves are planed and given the correct *shot* or bevel with a **Cooper's Jointer Plane.**

The largest of all planes, it can be as long as 6 feet or more. One end of the plane is supported on legs and the other rests on the ground. What is usually considered the bottom of the plane faces upward, and the wood is moved instead of the plane.

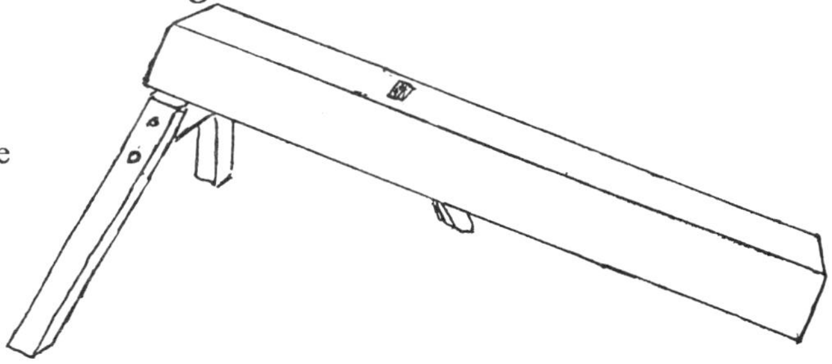

Step 2: Drawing the Staves Together into a Barrel Shape.

1. The staves are put together in a cone shape inside an iron hoop, and the staves are made more pliable using a **Cressett**.

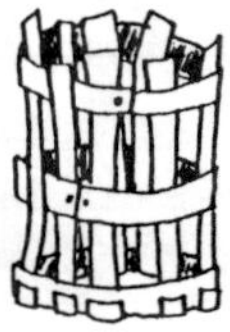

A container, usually made of old hoop iron, in which wood shavings can be burned. A cask is place over it to warm and make the staves easier to bend into the final shape. Water is also used to help produce steam.

2. The staves are drawn together by driving smaller and smaller **Truss Hoops** ...

A hoop of strong wood. A series of smaller and smaller hoops are driven down over the barrel to bend the staves together.

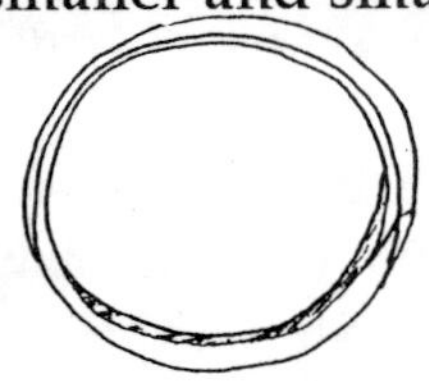

OR the staves are drawn together using a **Hooping Windlass**.

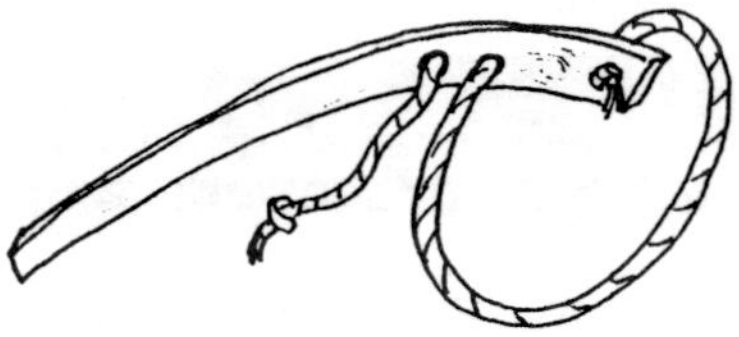

There are many hooping windlass designs. One uses a rope wrapped around the staves with a wooden lever to pull the rope tight, causing the staves to bend together.

Step 3: Riveting the Iron Hoops ...

1. A hole is put in one end of a metal hoop with a **Cooper's Punch** and **Cooper's Anvil**, and then the hoop is riveted together.

A steel bar with a blunt point at one end.

A T-shaped stake about 30 inches tall set in a block of wood. The top is slightly rounded with two holes to use with the punch. The anvil is also used to hammer over the hoop rivets.

OR Cutting, Joining and Nailing the Wooden Hoops.

2. The wooden hoop is cut and notched with a **Cooper's Notching Adze** OR a **Cooper's Whittle**.

Like a sharp adze except it may have a claw on the end or side. It is used to cut V-shaped notches in the ends of hazel rods to interlock to make a hoop.

A short, pointed knife with a blade about 2 to 3 inches long and a wooden handle.

3. If a Notching Adze is not used, hoops are nailed with a **Cooper's Nailing Adze.**

It is similar to other cooper's adzes except it has a cloven blade like a claw hammer. The poll has a checked surface for nail driving.

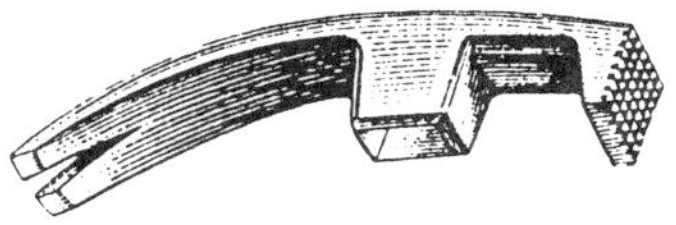

Step 4: Driving the Hoops onto the Barrel.

1. Hoops are driven on the barrel using a **Hoop Driver.**

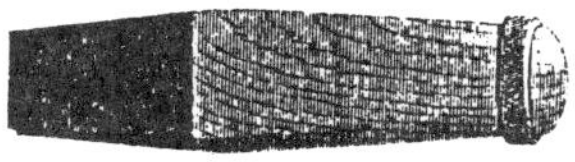

A wedge-shaped tool often with a groove on the nose to help keep the tool on the hoop. The wooden handle is often held together with a metal ring to prevent splitting while it is being hammered on.

Step 5: Cutting the Beveled Top or Chime.

1. The bevel or *chime* on the bottom and top of the barrel is rough cut with a **Cooper's Rounding Adze.**

Heavier and more sharply curved than other cooper's adzes. The blade is often ground at an angle which causes the blade to cut inward and helps stop the grain from breaking on the inside edge.

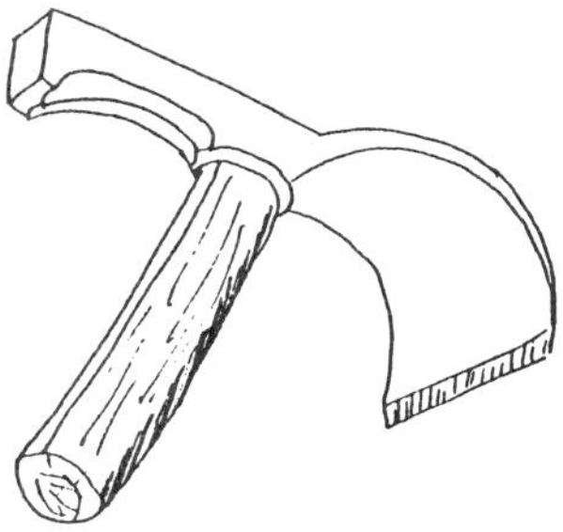

2. The chime is finished with a **Cooper's Sharp Adze.**

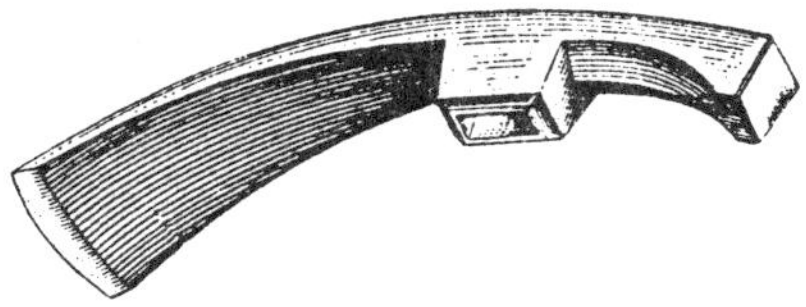

Like the rounding adze except smaller and lighter.

3. The top is leveled to leave a ledge for the howel and croze with a **Cooper's Sun Plane.**

A plane shaped like the arc of a circle. They are often made in pairs for right- and left-hand working.

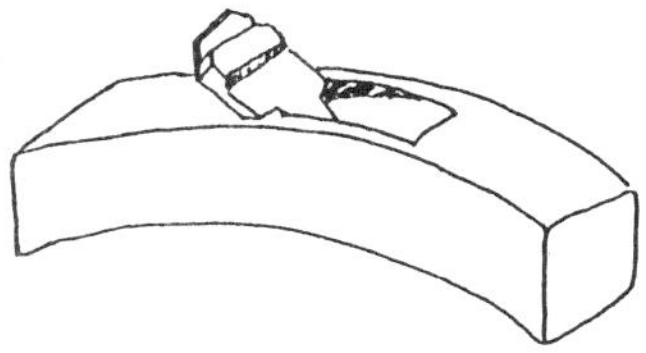

Step 6: Cutting the Howel and Croze.

1. The surface where the groove for the head will be cut is made flat with a **Cooper's Howel Plane.**

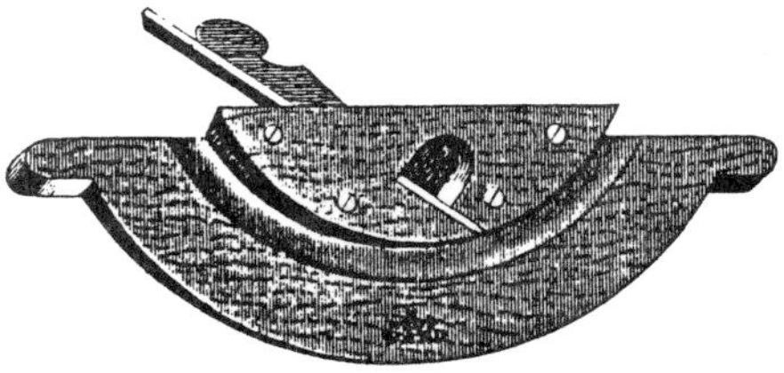

A plane with a convex cutting iron mounted on a circular fence

2. If the howel plane is not the right size, a **Cooper's Howeling Knife** or a **Cooper's Howel Adze** is used.

   A draw knife that has a short, bent, concave blade. The left handle is wooden and used outside the barrel. The right handle is an extension of the blade back and is used inside the barrel. Is considered a hard tool to use.

   A short, hollowed blade with a flat poll.

3. The croze groove to hold the head is cut into the howel with a **Cooper's Croze Plane.**

   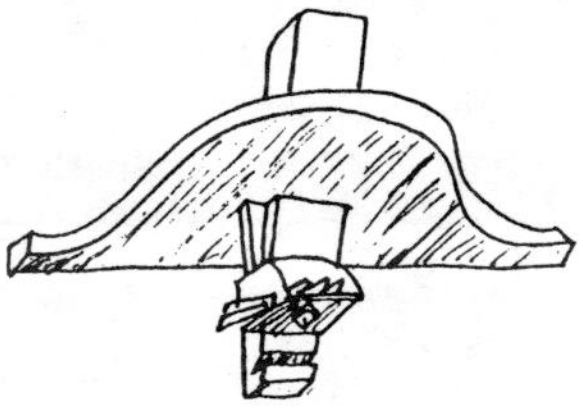

   A plow plane with a narrow cutter mounted on a stem which is fastened to a circular fence

Step 7: Cutting, Dowelling and Fitting the Heading in the Barrel.

1. The pieces or *cants* that make up the head are edged using the **Cooper's Jointer Plane** (see previous description).

2. Cant edges are drilled for dowels with a **Cooper's Brace.**

   A heavy drill brace with a large head so the cooper's chest can rest on it while one hand holds the head and the other turns the brace. A flat tanged spoon bit is often permanently attached.

   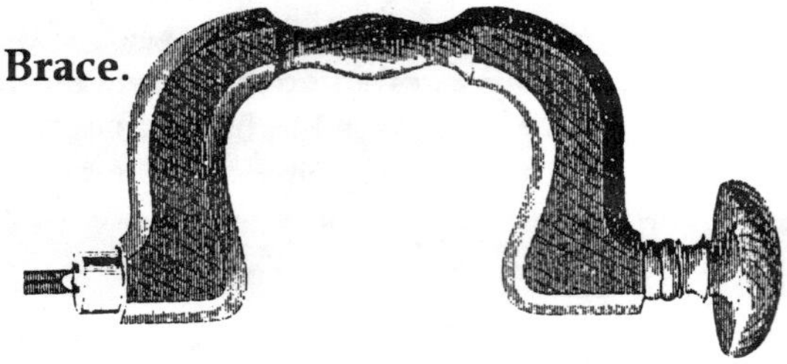

3. The pieces are put together and planed level with a **Cooper's Heading Swift Shave.**

   A shave that looks like a plane with handles that point up a little so hands will not be hit while working. The blade is slightly convex for working across the grain.

4. The circular shape of the head is marked with a **Cooper's Compass.**

   An iron or wooden compass up to 18 inches long.

   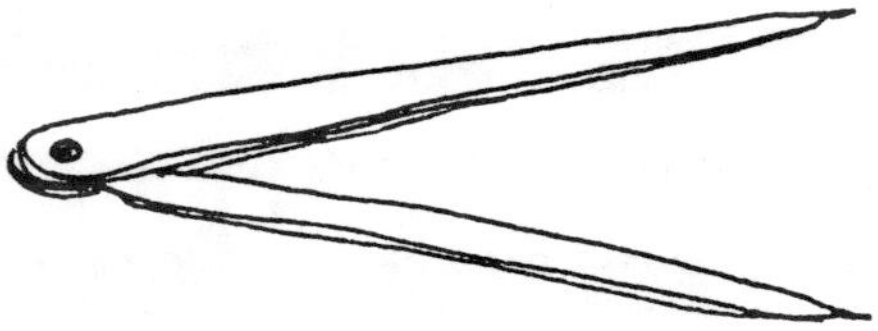

5. The circular head is cut with a **Cooper's Head Saw.**

   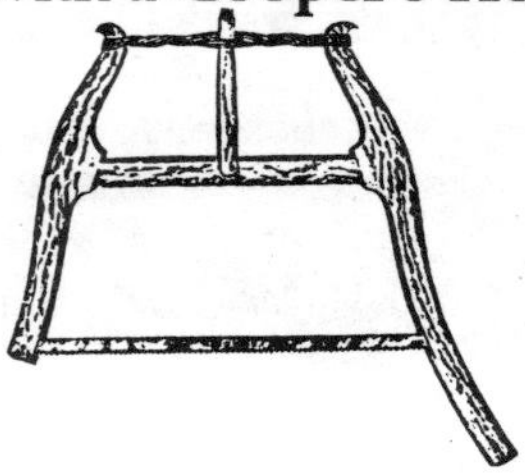

   A bow saw with a fixed blade at cannot be turned. The blade is about 22 inches long.

6. The head is roughly rounded and beveled with a **Cooper's Axe** (see previous description).

7. The bevel on the head is finished with a **Cooper's Heading Draw Knive**.

A large, flat draw knife up to 16 inches long.

8. Rushes are inserted into the croze of the barrel with a **Chincing Iron** and then the heading can be fit in.

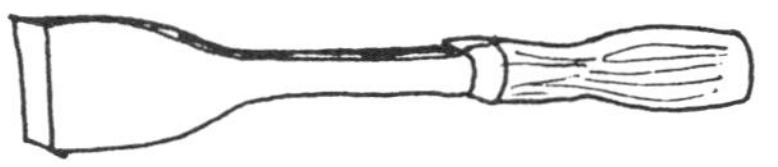

A chisel-shaped tool, either all metal or with a wooden handle.

Step 8: Finishing the Inside and Outside of the Barrel.

1. The inside joints between the staves are leveled with a **Cooper's Stoup Plane**, a **Cooper's Inside Shave** and/or a **Cooper's Round Shave Draw Knife**.

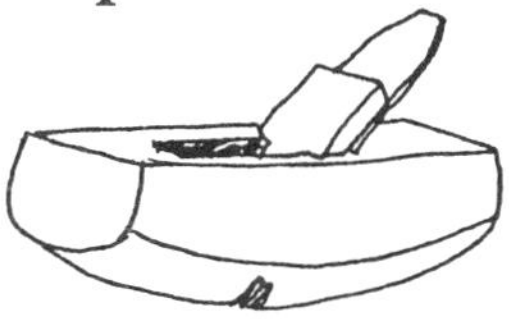

A plane with a rounded sole. It is made in different sizes for different size barrels.

A plane-like shave with convex sole and blade and up-turned handles. Made in different sizes.

A round blade in a complete or partial circle with a diameter of 2 to 6 inches.

2. The outside of the barrel is finished with a **Cooper's Downright Shave** and/or a **Cooper's Buzz Scraper**.

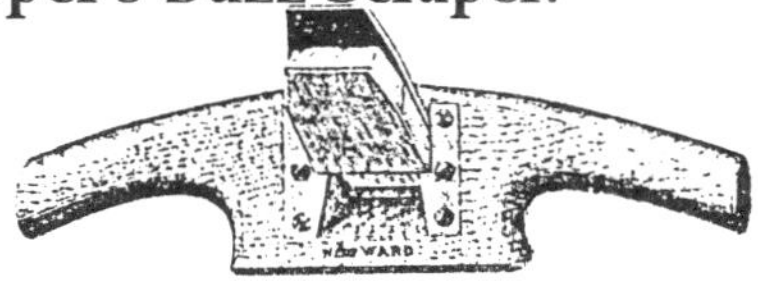

A plane-like shave with a straight blade. The handles are bent forward.

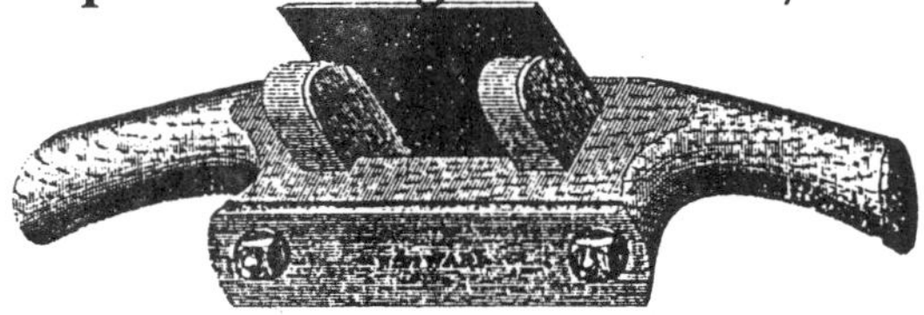

A scraper with a 3 to 5 inch blade set upright in a 14 inch stock. A larger, rougher version of the chairmaker's scraper.

Most, if not all, of the above steps were used by Dry Coopers to make Butter Tubs.

Making a Barrel With Machines

As suggested in a previous section, machines were invented to replace many of the hand tools mentioned in the last section. Machines could do just about every step in barrel making, especially for Dry Cooperage. Machinery designs could vary quite a bit - there are at least four different designs for cutting heading alone. Below are machine that could take barrel making from beginning to end.

1. Once the staves are rough sawn on a table saw, they are put through the **Machine for Dressing Rived and Sawed Staves.** This machine finishes both sides of the stave, inside and out, at the same time.

2. The edges of the staves are smoothed or jointed and given the correct bevel and bilge (or bulge) with a **Combined Fan and Stave Jointing Machine.** The fan blows any shavings away.

3. The finished staves are set around the bottom head with an iron truss hoop inside a **Barrel Setting-Up Form.**

4. A rope is wrapped around the flared end of the staves and a **Barrel Power Windlass** pulls the staves together and a truss hoop is put on.

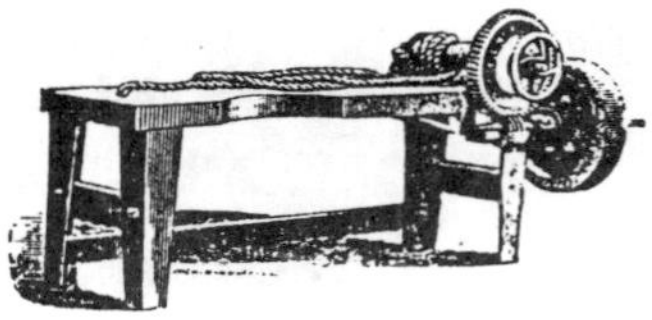

5. With the bottom and top truss hoop on, the rest of the hoops can be driven into place and then the barrel can be leveled so it will stand straight when upright. This could be done two different ways:

   a. A **Machine for Driving Truss Hoops** will drive the hoops into place, then a **Machine for Leveling Barrels and Casks** will straighten the barrel.

OR

   b. For Slack barrels, a **Combined Barrel Leveling and Truss Hoop Driving Machine** will both drive hoops and level the barrel at the same time. According to the manufacturer, this machine could do 2000 barrels in a day.

6. The leveled barrel is put on a **Chamfering, Howeling and Crozing Machine** which cuts the chamfer, howel and croze on both ends at the some time. This machine could do 1000 to 1500 barrels in a day.

7. The barrel is completed, but the heads still have to be cut. Heading blanks are cut and then the **Combined Fan and Head Jointing and Dowel Boring Machine** smooths the edges and drills the holes for the dowels. The fan carries all saw dust and shavings away.

8. Once the heading is put together with dowels by hand, the square blanks are put through a **Head Dressing and Leveling Machine**, which gives them a smooth and level surface. The machine could do from 5000 to 6000 blanks in a day.

9. The heads are then cut and the edges beveled by a **Machine for Turning Barrel Heads.** This machine can cut 5 tight head or 10 slack heads a minute, and can cut different sizes.

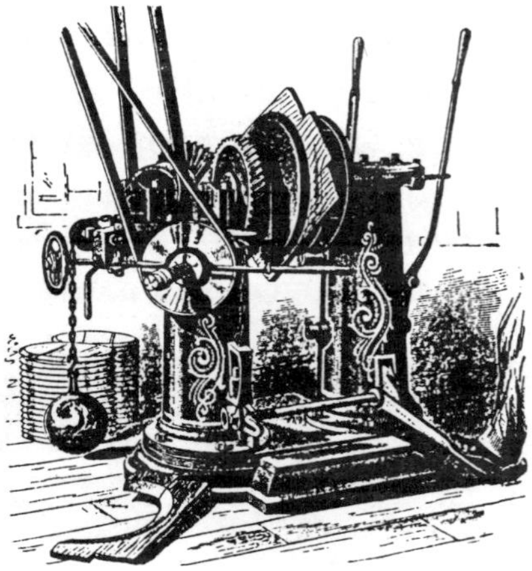

Who Made Barrel Making Machinery?

There were many companies that made barrel making machinery. The following is just one example of many.

The machinery in the proceeding pages and below were manufactured by E. & B. Holmes. The company was founded in 1854, and, at first, only made barrels. These barrels were used in the Great Lakes grain trade. It evolved into a manufacturer of both barrels and barrel making machinery.

The company started business in Buffalo, New York, but moved to South Buffalo in 1895. There, it could take advantage of the following:

1. The Buffalo River as a water source.
2. The New York Central Railroad yards for transportation.
3. The Ohio Basin of the Erie Canal for transportation.
4. Various South Buffalo iron works for materials.

E. & B. HOLMES,
Manufacturers of
Barrel, Keg, Shook and Stave
Machinery.
BUFFALO, N. Y.

E. & B. Holmes won awards for their machinery at the following exhibitions:

1. The World's Columbian Exposition in Chicago, IL - 1893
2. The Pan American Exposition in Buffalo, NY - 1901
3. The World's Fair in St. Louis, MO - 1904

Surprisingly enough, the company still exists today in South Buffalo, NY. It sells barrel making machinery to distillers in the Caribbean and repairs barrel making machines. The original South Buffalo factory, carpentry shops and shipping areas still exist. They still have a complete file of shop drawings and specifications, plus catalogues and photographs of their machines. However, the company will probably close when the present owner retires (perhaps within the next few years).

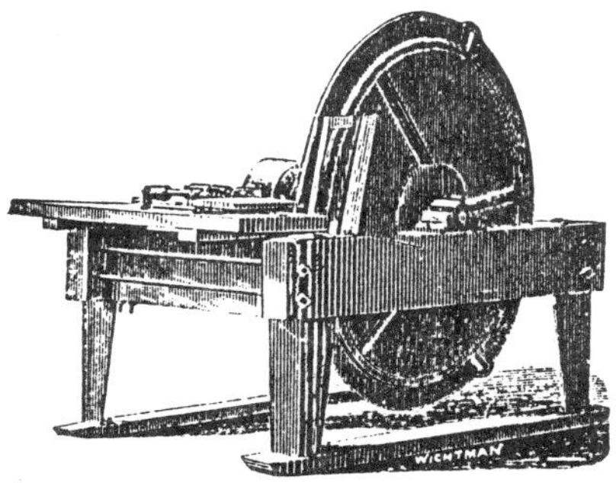

Machine for Dressing and Jointing Heading

Machine for Dressing Sawed Staves

A COOPERING FAMILY - The Harvey Barnes Clan

The Hanford order books have at least **119 different names** of people who ordered tub covers! We do not know wether they used hand tools or machines to make their butter tubs. Usually all we know about these people comes from their orders, their letters to the Hanfords and the Hanfords' answering letters. We do happen to know more about one particular cooper - Harvey Barnes of Oneonta, New York.

Harvey Barnes' Family Tree

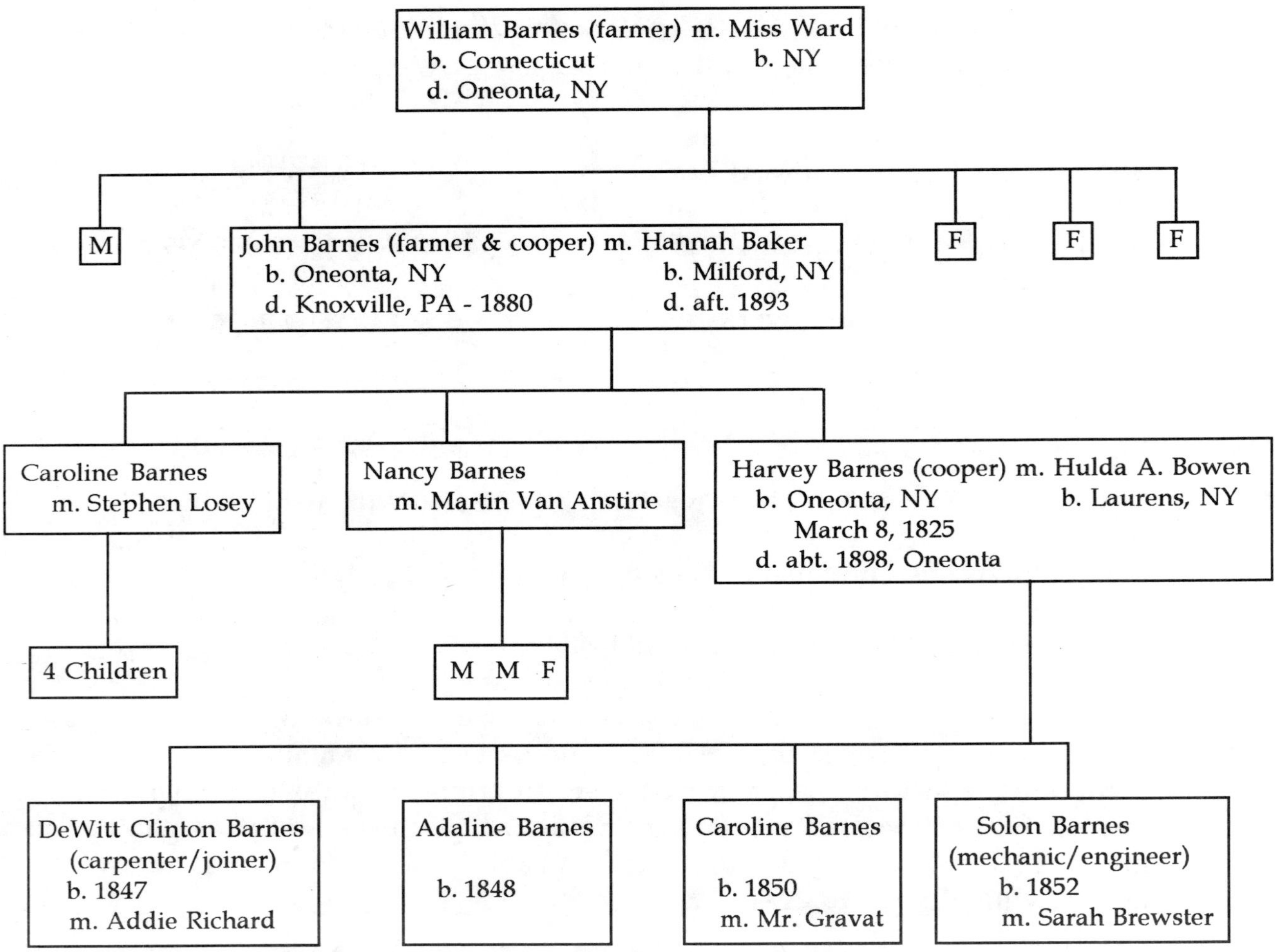

Of the Barnes family shown above,
four people ordered butter tub covers from the Hanford Mill:
Harvey, Dewitt Clinton, Adaline and Solon.

Harvey Barnes' Grandfather - William

William Barnes was known as one of the earliest settlers on the Susquehanna River in the Oneonta, New York area. He arrived some time in the late 1700s. He was known as a great lover of horses and knew quite a bit about them. William was a farmer. He was not in the cooper trade at all.

Harvey Barnes' Father - John

John Barnes started life as a farmer, but learned the coopering trade and became one of the first coopers in Oneonta, NY. A late 1800s book of Otsego County, NY biographies tells us this about John Barnes: "He was a great reader, and possessed a very retentive memory, having committed to memory the entire Scriptures, and being able to quote them with great accuracy. He was, however, somewhat heterodox in his views on religion, and became a strong believer in the universal salvation of the race. He was also opposed to secret societies, being an anti-Mason." John Barnes was also known as a Whig, but later switched his allegiance to the Republican party.

Harvey Barnes

The same biography book that describes John Barnes, said this about his son, Harvey: "... well known as a public-spirited man, and one of great personal generosity ..." He was a Democrat until Fremont's candidacy in 1856. At that point, Harvey switched to the Republican party. His cooperage was located at number 269 on Main Street in Oneonta. One source says it was powered by Silver Creek, but an 1868 atlas shows him on Elk Creek. According to his biography his shop was at the Main Street location from 1850 until his death. It was considered one of the landmarks of Main Street.

Harvey Barnes Cooperage Business

1. First and Last Order:
   Harvey Barnes made his first recorded order to the Hanfords on July 5, 1877. The last order from the Barnes Cooperage was made by Harvey's son, Solon. The last recorded order was made on March 19, 1900, two years after Harvey's death.
2. Did Harvey Barnes Always Order from the Hanfords?
   Since the information above indicates that Harvey had a cooperage from 1850, it would suggest that he did not always get his covers from the Hanfords. Perhaps he ordered them somewhere else, or more likely, he made them by hand. Between 1878 and 1886, Harvey may have found a better or cheaper supplier, for there are no recorded orders from him. Later letters suggest that Harvey knew a cover producer in Jamestown, NY. Perhaps that is where he bought his covers.
3. How Stable was Harvey Barnes' Business?
   From Hanford records, it appears Harvey had a stable business. He paid his bills on time (most often monthly). The Hanfords recorded no complaints. Harvey's orders to the Hanfords suggest he had a medium to large size operation. Below is a graph showing his tub cover orders for the year 1888:

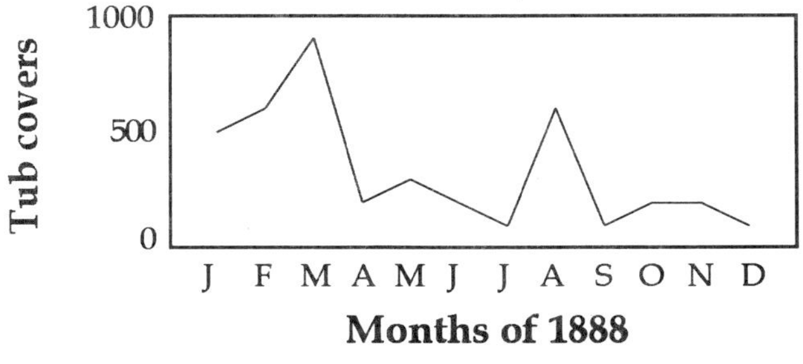

The Barnes Family in the Cooperage Business

Three of Harvey Barnes children were involved with his cooperage either when he was alive or after he died. Why did each of them start working with the business? Why did each of them eventually leave the business?:

DeWitt Clinton Barnes

In 1887 and 1888, the Hanford records show occasional orders from "D.C. Barnes" for "H. Barnes." This was DeWitt Clinton, Harvey Barnes' son. DeWitt was about 40 years old at the time. As far as Hanford records show, DeWitt Clinton never worked for his father again. In the 1893 biography book he was listed as a carpenter and joiner.

MYSTERY:

DeWitt Clinton was making orders for his father,
but what was his position in the Harvey Barnes Cooperage?

~ Was Harvey Barnes often ill in those years
so DeWitt Clinton had to take over running the business?
OR
~ Was DeWitt Clinton just one of the cooperage employees
told to make the order when his father was too busy or away on business?

Adaline Barnes

Some time between April 6 and June 3 of 1898, Harvey Barnes died. His daughter, Adaline, took over the running of the cooperage. She ran the business for a little under a year, and then she no longer appears in Hanford order records. She was about 50 years old at the time and unmarried. She did leave us with some memorable letters:

Oneonta June 17, 1898

Mr Hanford

Dear Sirs

I never saw such poor covers in all my life you never sent Father such trash the farmers and custumars are kicking some of them will not take them from us now dont think for one moment that because you are deeling with a woman that we dont know good covers I did not see them untill to day my cooper said he never saw you send out such poor stuff he is a good Judg he has worked for my Father most 40 years now I think you had ought to make it right Some of them are not evan plained on one side and I could put them to gather better my self now if you send us such covers we will have to send to Jamestown for them let us hear from you

Respectfully yours
Adaline Barnes

Oneonta June 28, 1898

Mr Hanford

Dear Sir

Can you send us 200 more covers rite away as we have them most used up our cooper said those others were some of them quite poor send good ones the farmers are quite particular with us it is not as tho our Father was living they think we do not understand it as well but we have the same cooper that Father had for 40 years or more let me hear from you by return of mail and oblige

Mis A Barnes

Mr Hanford can you tell us where we can get some Hickery Soap pales if so let us hear from you by return of mail and Oblige

Adaline Barnes

Adaline Barnes seems to have calmed down from her first letter. The quote "...such trash..." from the first letter is much harsher than "...some of them quite poor..." from the second letter. There were other complaint letters sent to the Hanfords, but none quite as strongly worded as Adaline's. It appears that the Hanfords always tried to make things right. Unfortunately, we have no letters from the Hanfords answering Adaline's problem. We do know she continued to order from the Hanfords.

MYSTERY:

In those years, it would have been unusual for a woman, a fifty year old spinster, to take over the running of an industrial shop or factory.
There are quite a few questions that surround the year Adaline Barnes worked for the cooperage:

*~ Did Adaline take over her father's business or was she just the secretary?
The way she writes suggests that she did take over the cooperage operations.
Why didn't one of her brothers step in right away?*

*~ Were the covers the Hanfords sent actually inferior covers
because they knew that Adaline was in charge?
OR Were Adaline and her cooper more picky than her father, Harvey, had been?
OR Did the shipment just happen to be a fluke,
a mistake that could have been sent to any customer?*

*~ Why did Adaline hand the business over to her brother, Solon,
who was a railroad engineer and not a cooper or woodworker at all?
Was it always understood that Adaline would only run the shop until a brother could take over
or was she forced out by her brother?
Did Adaline enjoy running the cooperage?*

Solon Barnes

By February 1899, Harvey Barnes' son, Solon, was making the cover orders for his father's cooperage. Solon was a mechanic and engineer with the railroad in Oneonta, but it appears that he replaced his sister in the operation of the cooperage. Solon's last existing order was made on March 19, 1900.

Mystery:

Why did Solon Barnes stop ordering tub covers from the Hanfords?

~ Did Solon find another cover supplier?
OR
~ Did he close his father's shop or just the butter tub portion of the business?

If Solon did close the shop, why did he do it?

~ Perhaps, Solon closed the business because he was not interested in it. His training was not in woodworking, but with steam engines and the railroad.
OR
~ Another possibility was the growth of creamery and milk bottling plants that cut the demand for butter tubs (see Chapter 4 - The End).

Other Hanford Customers - Payment Problems

While Harvey Barnes always paid his bills on time, not every Hanford customer did. Below are two somewhat humorous letters dealing with payment problems:

1. This first letter is written by the Hanfords to C.H. Brady. Notice that even though the Hanfords are having problems collecting payment for their bill, they still plan to send Mr. Brady his next cover order.

C.H. Brady
Delhi, N.Y. *July 31, 1902*

Dear Sir:--

We are in receipt of yours 30th ins't enclosing check $10.00 on account. We note what you say in regard to your Father not being able to get money to send us, and wish to say in this connection that we are not selling these goods to your Father, but to you. We enclose a statement of a/c to date for which we would like to have a check by middle of next week.

Will send some covers and heads soon.

Yours respectfully
Hanford Bros.

2. This letter is from the wife of a Hanford customer.

New Kingston, N.Y.
D.J. Hanford & Sons *Nov. 20th 1903.*
Messrs.
Your bill resed to day which is the second one with in a week. I think you are unnecessarily frightened. Mr Douglas is absent from home on buisness since before the rect of you bill a week ago but will be home Sat. night I think. I never send checks during his absence without him asking me to do so. I can assure you you will get your pay and I think after as much square dealing as Mr Douglas has had with you you ought to have some confidence in him your self that he is not a skin. although he may have been painted to you as such. I think you will have yet to find a debt he has failed to pay. On his arrival home I shall certainly not forget to present your letter that he may attend to the matter.

Mrs Geo. Douglass.

Mystery:

George Douglass was a good customer, so why did the Hanfords send two bills in a week?

~ Was it an oversight? Did Mrs. Douglass overreact?

~ Was someone else "painting" George Douglass as a "skin" to the Hanfords?

OTHER HANFORD CUSTOMERS - Reputations

It is interesting to note that two years before Mrs. George Douglas is worried about her husband being "painted" as a "skin" (see previous page), her husband is doing the same to someone else's reputation:

Geo. Douglas.
Manufacturer of and Dealer in
Tubs, Firkins, Barrels and Churns.

New Kingston, N.Y. Dec. 18, 1901

D.J. Hanford & Sons
Gents
Inclosed pleas find Check to bal my account. as a friend, I will take the liberty to inform you in regard to C.W. Hallock. would say unless you get cash or security you had better keep your goods. Just ask W.C. Johnston, the back smith in regard to him
Thanking you for all fast favors I remain yours
Truly Geo Douglas

There is no evidence whatsoever in the Hanford order books or in their letters to indicate that C.W. Hallock was ever a problem. He always paid his bills on time. Mr. Hallock made orders until at least 1909.

MYSTERY:

Why did George Douglas write this letter,
especially if C.W. Hallock actually was a good customer?

~ Did the Hanfords write to George Douglas
to ask his opinion of C.W. Hallock (they did in other cases)?
OR
~ Was George's letter just an unsolicited warning?

Other Hanford Customers - Mix-ups & Mistakes

Sometimes mistakes were made when the Hanfords filled butter tub cover orders. The Hanfords always seemed to find someone else to blame, but they always attempted to fix the problem. Here are a few examples from the letters in the museum archives.

1. To this date, there is only one case where the Hanford Mills Museum archives have both the complaint letter and the Hanford's answer to the situation:

> *June 30 1902*
>
> *Roxbury*
> *Hanford & Son*
> *received heads and find them only 12 inches instead of 12 1/8 as Ordered make the rest 12 1/8 now could you make the heads cheaper if matched it seems to me so and I would just as soon have them as they are for cheap tubs can you make what covers & heads I need? ...*
>
> *Yours G.S. Dudley*

> *G.S. Dudley*
> *Roxbury, N.Y.* — *July 1, 1902.*
>
> *Dear Sir:--*
>
> *We regret to learn that the heads were not as ordered with reference to size; we talked with the man who turned these heads and he claims they were 12 $^1/_8$", but we think he might possibly be mistaken. We are turning some more for you to day which we are sure will be all right.*
>
> *As to matching heads, would say, we will make them in this way for $3.50 per hundred. Please let us know if we shall make them this way hearafter.*
>
> *We can furnish you what covers and heads you may need and will be pleased to do so.*
>
> *Yours respectfully,*
> *Hanford Bros.*

2. George Douglas (mentioned in the previous sections) wrote with a complaint in 1902. Unfortunately, his letter does not exist in the museum archives. All that remains in the Hanford's answer.

> Geo Douglas, July 15, 1902
> New Kingston, N.Y.
>
> Dear Sir:--
> Your order for 500 Maple heads received. Same will receive prompt attention - will ship at once, 2 Bbls Ash and Bbl Maple and will ship balance later. Let us know at once if you can use the ash instead of maple for balance of order. We regret to learn that you have received some maple heads which were not satisfactory. It is an oversight on the part of the man who packs the heads if any are sent as you state, and we are willing to allow you for them if they are unfit for use.
>
> Yours respectfully,
> Hanford Bros.

3. Perhaps the most interesting mix-up involved both George Dudley and George Douglas. The situation is not surprising considering the similarity of the names. One might wonder, though, why George Dudley and George Douglas always have all the problems. It is probably just coincidence. It is too bad some of the letters are missing.

> G.S. Dudley, Sep. 12, 1902.
> Roxbury, N.Y.
>
> Dear Sir:--
> On Aug. 8th we shipped some heads to you and some to Geo. Douglas Arkville, N.Y. and in some way they have become mixed up. Douglas claims he has one Bbl. of heads 12 1/2" and we think you probably received one of his Bbls of 12-5/8" heads; if this is so, will you please ship the bbl. to Geo. Douglas Arkville, N.Y. and we will send him an order to ship the bbl. of 12 1/2 to you.
>
> If you are to any extra expense in this matter, please let us know how much it is and we will credit your account with same
>
> Yours respectfully,
> D.J. Hanford & Sons.

3. The Dudley/Douglas Mix-up continues with another letter written on the same day as the one on the previous page:

Geo. Douglas, *Sep. 12, 1902.*
New Kingston, N.Y.

Dear Sir:--
We are in receipt of yours ins't with reference to the heads which you received, and which were intended for G.S. Dudley.
Please ship this bbl. of heads to G.S. Dudley, Roxbury, N.Y. We have written Mr. Dudley to ship the bbl. of 12-$^5/_8$" heads to you if he has them on hand, and if he does not do it we will credit your account with the ones you ship to him. Please let us know if we made this mistake or was it the R.R. Co. We note your order for covers and heads and will ship some time next week.

Yours resepectfully
D.J. Hanford & Sons.

In this letter, it appears that the Hanfords may be getting a little testy:

Geo. Douglas, *Sep. 18, 1902.*
New Kignston, N.Y.

Dear Sir:--
We have received word from G.S. Dudley, Roxbury that he has used some of the heads which he received through error of R.R. Co which were intended for you. Please take the barrel which was marked for G.S. Dudley, Roxbury, N.Y. to the Arkville station and call the agent's attention to the error and have him send it where it belongs. Send us the number of heads it contains and we will credit your account with same. You should have refused this barrel when you discovered the wrong address on it. Please attend to this promptly, and oblige

Yours respectfully,
D.J. Hanford & Sons.

3. And then the Hanfords wrote to the cause of the mix-up:

Freight Agent, *Sep. 18, 1902*
Arkville, N.Y.

Dear Sir:--

We understand from Geo. Douglas, New Kingston, N.Y. that you delivered to him some time in August this year a barrel of heading which was marked G.S. Dudley, Roxbury, N.Y. As this heading is not suitable for Mr. Douglas' use, we have instructed him to return the barrel to you to be placed at proper destination. Mr. Douglas will undoubtedly make claim upon us for freight and other charges which you will please attend to, otherwise, we will have to report the matter and make claim to the R.R. Co. Please give this prompt attention, and oblige

Yours respectfully,
D.J. Hanford & Sons.

Below, you will find a map that shows where the Ulster and Delaware Railroad ran, the locations of Roxbury, New Kingston and Arkville, and where these places were in relation to East Meredith:

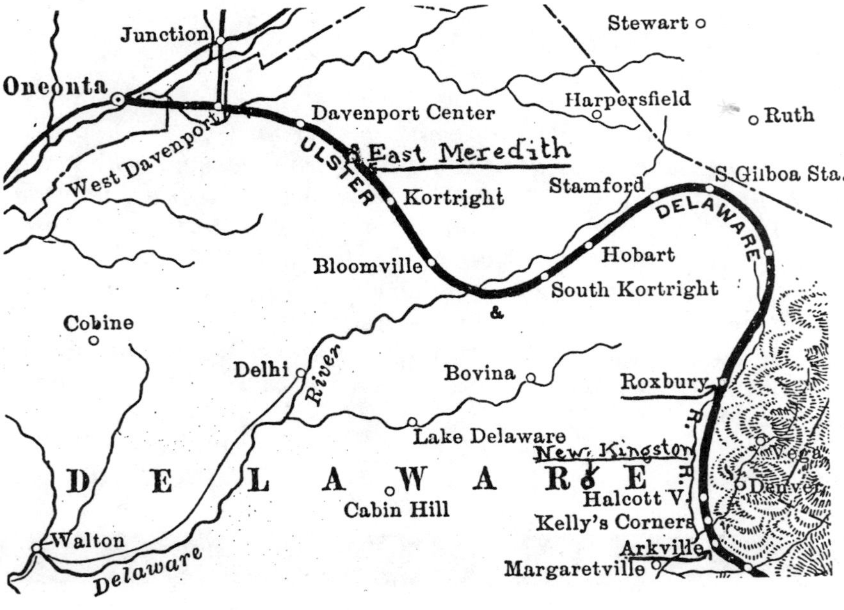

George W. Sides - At the Beginning & the End

George W. Sides, of Schenevus, NY, was perhaps the Hanfords' most persistent customer, though he couldn't always be called their best customer. To this date, little outside research has been done into his personal or business life. However, buried in the Hanford business records, there are some indications of what George Sides was like and how his business was run.

George Sides made his first recorded order to the Hanfords in April 1878, about a year and a half after they first started production. George was also the last person to place a tub cover order on April 6, 1917 (which the Hanfords could not fill). There were times when he did not make orders, such as from 1880 to 1886. Perhaps George Sides found another mill to order from, or maybe he went out of business for awhile, or it is also possible that the Hanford records were not good for that period (since records also show that the Barnes family did not order in that period).

In 1879, George Sides was doing business with a partner and signed all his letters, "Sides & Wolcott." Again, in 1904, he had another partner and letters came signed, "Sides & Youmans." In each case, the partner was gone within a year or two.

Mystery:

Why did George Sides have so many partners?

Did he need partners for financial help?
OR
Did he take partners for other reasons? What would those reasons be?

Was George Sides hard to work with in some way so that his partnerships did not last long?

Unlike the Barnes family, George W. Sides made small orders less frequently. Below is a graph showing George's banner year, when he bought more tub covers than he ever did before or after. The pattern of purchases is totally different than Harvey Barnes'.

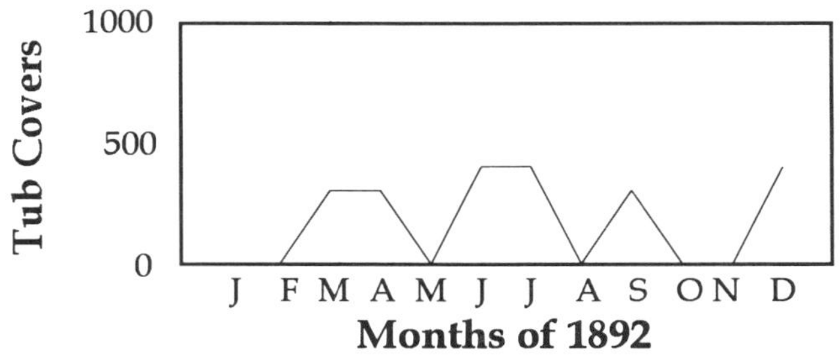

In 1898, George wrote the Hanfords explaining why he had difficulty paying his bill. It was a hard-luck story, but he offered to pay the Hanfords with butter tubs. This payment method is often offered in later letters. Hanford order books and other records indicate that, at least after the turn-of-the-century, George Sides was expected to prepay for all orders - perhaps to avoid these payment problems. Below is George's 1898 letter. It may help to read his letters out loud, since he seemed to spell phonetically. George's spelling may also give indications of how he actually spoke, as when he uses the word "extrey."

Schenevus *Nov 19/98*

Mr D.J. Hanford Dear Sir your leter receaved I dont blame you for demanding your pay it is not because I meant to cheat you that I havent paid you before it is due to the lose of my fingers to years ago I got behind being lad up so long the summer of 1897 was a compleate failure with me I lost $150(?) in a deal with one man thet year I have done well but when I got a dollar I have had three places for it so I start in the winter with no money I dont see how it is posible for me to rase the money for you at the present time cant you take tubs enough for me to pay the bill I will deliver them to you for 25¢ & you furnish covers that will make them 32¢ thare that will give you a chance to make a profit on the tubs & will help me out but I cant deliver the tubs before the fore part of Jan my stock isent dry enough to work yet I bought the timber & sawed it my self I hav some extrey fine stock I will guarentee every tub to be right there are what we call spring tubs for quick shiping

Yours Truly
Geo W Sides

A little over one year later, George W. Sides again had problems paying his bills. Again, he offered tubs in payment, and blamed his problems on the "milk station" or creamery this time - a valid enough reason (see Chapter 4 - The End).

Schenevus *Jan 9 / 1900*

D.J. Hanford & Sons
can I let you have some tubs for covers I need some covers & dont get enny money this time of year since the milk station has been here I dont get mutch fall & winter trad I want some covers on hand when the trade does start up it will be the last of next month on the first of March before I begin to sell I can sell all I can get out when spring opens I can give you a good tub my stock is fine I can let you have tubs for 30¢ & you find covers we are going to sell 5# higher this season I in tend to use more covers this season than I did last if you will take some tubs for covers I will bring some over next weak to you if the roads is good let me no soon

Yours Truly
Geo. W. Sides

By the next year, George Sides was obviously asked to prepay his orders. In this letter he is sending a prepayment, but also notice that he still owes for a previous account.

Schenevus
May 24 / 1901

D.J. Hanford & Sons
please find check for $8 Eight Dollars please send me 200 covers 16 1/2 I told you I would pay you some on the old bill when I ordered agane but trad is Just starting up so I dont have enny more at presant then I am oblige to have for stock nex month I will try & balence the account with you

Yours Truley
Geo. W. Sides

George W. Sides continued as a Hanford customer. Finally, in 1917, nearly a whole year after his last order, George became the last person on record to order tub covers from the Hanfords.

Schenevus
Apr 6 / 1917

Hanford Bros *Dear Sirs*
would it bee enney indusement to you to make me some covers if I would pay you $5.00 per 100 I will do that if you will make me some covers this spring & summer

Yours Truley
Geo. W. Sides

April 10 17

Mr. Geo. W. Sides,
Schenevus, N.Y.

Dear Sir:--

Replying to your favor of the 6th, instant beg to say that we cannot make any covers from the fact that we have no lumber in stock to make them of. The Tub Cover business got so poor that we discontinued getting out stock a long time ago and have rearranged our factory for other work. Regretting our inablility to serv you, we are

Respectfully,
HANFORD BROS.

The Hanfords sent back this letter in response.

And George, perhaps not seeing the end of the butter tub era, sent this letter to the Hanfords.

Schenevus, NY
Apr 23 / 1917

Hanford Brothers
if you are closing out the tub cover buisness what is the lowest you will take for your matcher I havent got a good matcher give me your bottom price on it if you want to sell write me by return mail

Yours Truley
Geo. W. Side

There are no records that indicate whether the Hanfords sold George a matcher or not. Hanford Mills Museum still owns an original Hanford matching machine (see Chapter 1 - The Mill).

MYSTERY:

Did the Hanfords sell George W. Sides a matcher?

Did they sell a more up-to-date matcher that there are no surviving records of, and keep the older matcher (which the museum now uses)?
OR
If they didn't sell their one and only matcher, why didn't they?
Did they want to keep it to use if they needed it?
Did George have enough money to meet the Hanfords' sale price?

Chapter 3

The Farmer:

THE COW - What is Milk?

Today, most people don't think beyond the cardboard carton or plastic jug that holds their milk. Milk doesn't come from a jug or box, it comes from an animal. Through history many animals have been used to provide humans with milk: camels, cows, goats, horses, llamas, reindeer, sheep, water buffalo and zebu. In the East Meredith of the 1800s, though, milk came almost exclusively from cows. The cow provided milk, and from that milk, farmers made products like butter, cheese, and other items.

So, what is milk?

Water	87.59	%
Butterfat	3.67	
Casein	2.63	
Albumin	.60	
Lactose	4.78	
Ash	.73	

Milk is mostly water, but it is its minor ingredients that make milk what it is.

~ **Butterfat**: Butterfat is animal fat. It can be separated from milk and used to make butter. What is left is known as skim milk. Usually, not all the butterfat is removed from milk. When you buy 1 or 2 percent milk in the store, the number is referring to the butterfat content. You will notice, even "skim milk" has a percentage of butterfat.

~ **Casein**: Casein is a protein found only in milk. It is often called Calcium Caseinate because it is found in combination with calcium in milk. It can only be separated from milk with special equipment. Casein is used in the production of many products, including plastic, glues, paints and foods.

~ **Albumin**: Albumin is also a protein that can be coagulated using heat - this is how ricotta cheese is made.

~ **Lactose**: Lactose is sugar. In humans, two enzymes control lactose digestion. One is present in almost every human at birth. This enzyme allows them to digest milk for the first few years of their life only. The second enzyme allows adults to digest lactose. Most of the world's adults do not have this enzyme, and so they cannot digest lactose. Some genetic groups (including people descended from northern European and certain Nigerian areas) do have the second enzyme and so can digest lactose and milk.

~ **Ash**: Ash is mineral salts such as iodine, potassium, sodium, magnesium, calcium, etc. Milk can be an important source of many minerals required in the human diet.

While the typical 1800s East Meredith farmer may not have understood all these ingredients, dairy scientists in the late 1800s did. This book focuses on the butterfat ingredient in milk. All dairy people in the 1800s knew about this ingredient.

The Cow - Where Milk Comes From

A cow gives her first milk when she has her first calf, usually when she is about two and a half years old. Several days before her calf is born, the cow starts to secrete a special milk known as colostrum. Colostrum is high in antibodies and vitamins, and is vital for the calf's survival. After about four or five days, the cows begins to produce what we consider milk. The whole process is known as "freshening." When a cow has "freshened," she is then available to milk.

Today, a cow's lactation period (the period of time in which she gives milk) lasts about 300 days. Elizabeth Hanford's diaries indicate that their cows in the late 1800s may have had a lactation period of only 210 to 270 days. Top producers may have reached today's 300 day lactation period.

A cow can be milked even after she is pregnant again, but will stop giving milk two to four months before her next calf is born. This is known as "drying up." Cows remain good milkers for about five years after their first lactation period, though it is not unusual to find cows who can be milked ten to twelve years.

The cow usually has an udder made up of four glands, divided into a fore- and hind-udder. Each of these quarters usually has one teat, though they can sometimes have more. Milk is formed in the udder and secreted through the teats.

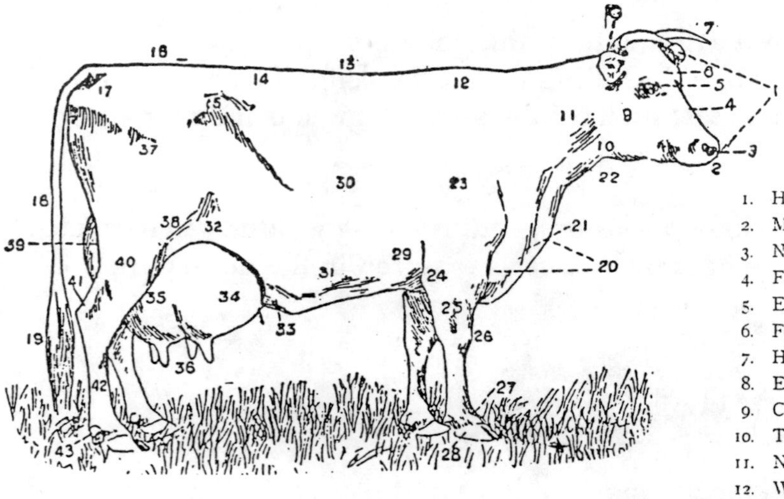

1. Head.
2. Muzzle.
3. Nostril.
4. Face.
5. Eye.
6. Forehead.
7. Horn.
8. Ear.
9. Cheek.
10. Throat.
11. Neck.
12. Withers.
13. Back.
14. Loins.
15. Hip bone.
16. Pelvic arch.
17. Rump.
18. Tail.
19. Switch.
20. Chest.
21. Brisket.
22. Dewlap.
23. Shoulder.
24. Elbow.
25. Forearm.
26. Knee.
27. Ankle.
28. Hoof.
29. Heart girth.
30. Side or barrel.
31. Belly.
32. Flank.
33. Milk Vein.
34. Fore Udder.
35. Hind Udder.
36. Teats.
37. Upper Thigh.
38. Stifle.
39. Twist.
40. Leg or gaskin.
41. Hock.
42. Shank.
43. Dew claw.

THE COW - Milking

Cows should be milked at least two times a day. Some argue that more frequent milkings are better, explaining that cows milked often gave more milk. Most farmers do two milkings as the most convenient, once in the early morning before breakfast and once in the evening usually before supper.

How do you get milk from a cow?

1. Cows are creatures of habit. They like to be milked at the same time, in the same place, with the same equipment, by the same person. If the routine is familiar, the cow "lets down" her milk and milking will go easily. If anything changes, the cow may not give her usual amount of milk, and milking will be difficult.

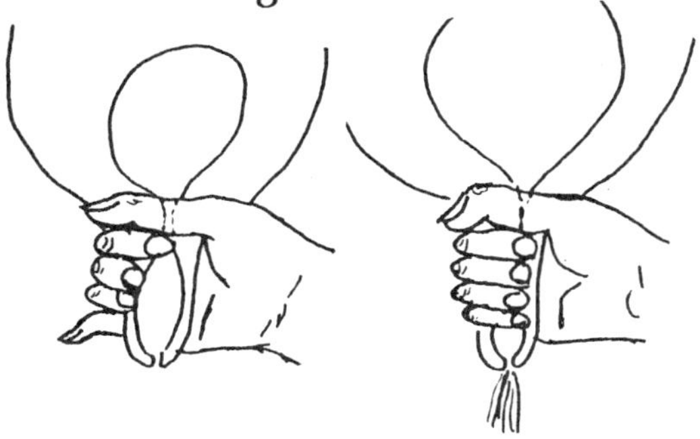

2. Before milking, clean the udder, then gently grasp the teat near the udder with your thumb and first finger. Use each successive finger to press the milk from the teat into a container. Then let the teat fill and repeat the operation using both hands alternating on two teats. Then move to the next two.

3. The first milk drawn from each teat should be directed into a "strip cup" with a black interior or a strainer. This will help the person milking the cow to check the milk for any clumping or strange separation that may be an indication of disease. The first squirts of milk also have a high concentration of bacteria from the teat channel.

4. Once you cannot remove milk using the above method, the udder must be "stripped" by messaging each quarter with one hand while removing most of the last of the milk. The milk should be cooled immediately, ideally to about 40°F which will stop the growth of bacteria.

The milker has from four to seven minutes to milk and strip a cow. After that period, the cow's "let down" reflex subsides, and then it will be difficult to get any more milk from her. Any milk left in the udder will be reabsorbed. If the cow reabsorbs large amounts of milk she may "dry up" too soon, so it is important to get the maximum amount of milk before her "let down" reflex is over.

The Cow - Breeds

It is difficult, if not completely impossible, to say exactly what breed of cow each East Meredith farmer was raising. We do know that the breeds changed through history, and many farmers may not have had full-breed cattle at all. Unfortunately, most newspaper accounts, diaries, probate inventories, etc. mention only that there were cows on farms. Rarely do these sources mention specific breeds. The best source is the "For Sale" ads in the paper. These ads give the names of some people and at least provide a list of breeds found in the area. These were the cattle breeds available around East Meredith in the late 1800s and early 1900s according to the local paper and the memories of residents:

Holstein: This is the cow most often seen in dairy regions. It is originally from Holland, but this breed has been in North America for a long time. It usually is white with black spots.

Jersey: This dairy cow comes from the Channel Islands between France and England. The breed tends to be smaller than most other dairy breeds. These cows are known for their fawn color with black highlights.

Ayrshire: This cow breed originally came from Scotland and was known for being able to produce milk on sparse pastures. They tend to be white with red/ brown spots.

Guernsey: This is another dairy cow from the Channel Islands. Guernseys are mostly spotted light brown with white.

Devon: This is a dual purpose breed, meaning they were used for both beef and dairy production. They originated in England and are a red color.

Durham: This is breed is also called Shorthorn. They were most often used for beef production and were very popular until the mid-1800s. They are originally from England, and were either all red, all white or a shade of red roan between.

Hereford: This is another beef breed. Herefords originated in England, and were not common in North America until the late 1800s. They usually are red with a white face and belly.

Farmers interested in producing milk, butter, or cheese would have chosen cows related to the four dairy breeds mentioned on the last page. The dairy breeds can be divided further between cows that produced a large quantity of low butterfat milk and cows that produced a small amount of milk that was high in butterfat.

To make butter, farmers preferred the breeds that produced the milk high in butterfat. The best known breed in the East Meredith area was the Jersey. Today, since low-fat liquid milk is in highest demand, farmers have switched to the Holstein breed. The differences between these two breeds can best be understood when you read this humorous argument between breeders of Jerseys and Holsteins.

A JERSEY gives:
- a low amount of thick milk
- high in butterfat

A HOLSTEIN gives:
- a large amount of thin milk
- low in butterfat

VS.

The Holstein farmer complained about how little milk the Jersey gave, saying if you dropped a silver dollar into a pail of Jersey milk, the milk wouldn't even cover the coin.

The Jersey farmer complained about the Holstein's thin milk, saying that if you dropped a silver dollar into a pail of Holstein milk, you could see the coin right through all the milk.

How to Make Butter

Over the years, many tools and machines have been developed to help make, process and pack butter. These will be discussed in depth in the following section. The process of making cream into butter, though, remains relatively constant. No special equipment is necessarily required, but two things should be remembered:

1. Everything must be kept absolutely clean and sterile to avoid flavor and bacteria contamination. In the 1800s, the causes of such contamination were not completely understood, but it was noticed that anyone who kept a clean work area and tools made better butter.

2. 60°F is usually considered the best temperature at which to make and work with butter. Some farm manuals and advice books of the 1800s may have given a different temperature, but a good average was 60°. The actual temperature might vary depending on the breed of cow, the time of year, what the cow ate, or any number of other factors.

Separate

First the cream, which contains most of the butterfat, must be separated from the whole milk. This can be done a number of ways, including letting it sit until the cream rises or using a centrifugal separating machine. (NOTE: If you wish to make you own butter at home, this step is already done for you when you buy heavy cream at the store. You cannot separate cream from homogenized pasteurized milk from the store. Cream will only separate from fresh and unaltered milk from the cow.)

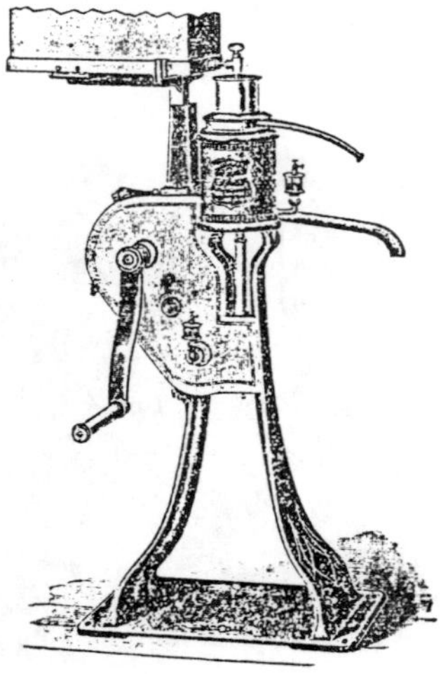

Churn

Put cream in a churn (even a simple glass jar can be used), and agitate steadily until the butterfat "breaks" from the cream or forms into a mass leaving a watery substance known as buttermilk (which is not the same as "cultured buttermilk" found in grocery stores today). Churning may take up to 45 minutes to do. If the resulting butter comes from a grass-fed cow whose milk carries carotene, the butter will be yellow. Jerseys and Guernseys are known for their yellow butter. If the milk did not have carotene because of the cow breed or the feed it ate, butter color can be added when the butter is worked.

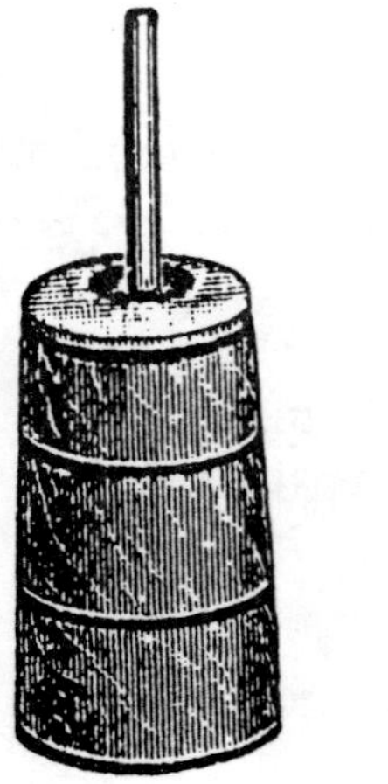

Wash

Remove the buttermilk from the churn (drink it, use it in cooking or feed it to farm animals like pigs). Pour cool, clear water into the churn and continue to agitate to wash the butter. Pour off water and repeat until water runs clear. Most suggest this should take no more than three washings. Butter can also be washed in a bowl instead of in the churn.

Work and Salt

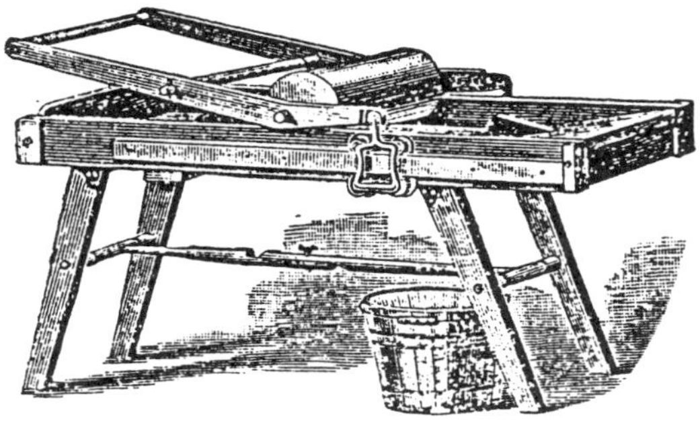

Put the butter in a bowl or on a flat surface and use your hands or paddles, and knead the butter to remove any excess water or buttermilk. Any water left in the butter will cause the butter to sour or turn moldy. There were also machines known as butter workers that worked butter. Once all the moisture is removed, a little salt and/or butter color can be worked in. Salt has to be completely dissolved and uniformly distributed throughout the butter. At this point, the butter is ready to eat, but in the past, if it was going to be shipped or sold, the last step took place.

Pack

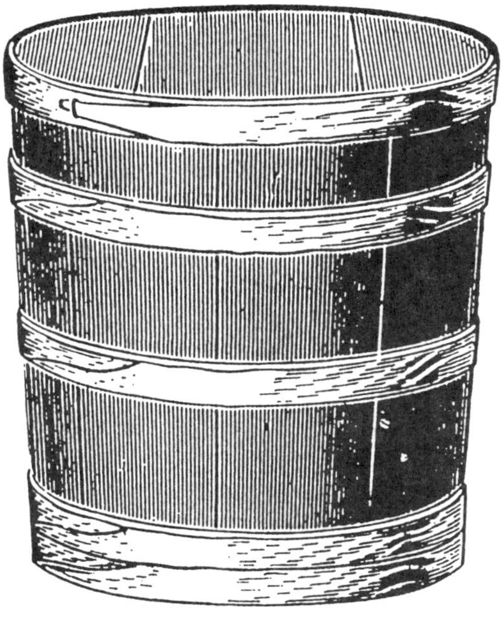

Prepare a butter tub or firkin. This is done by putting the tub through a series of hot water and hot brine (water and salt) soakings - guides and manuals differed on the exact procedure. This removes any wood taste and bacterial infestations. Last of all the tub is filled with cool, clean water to chill it. Sometimes the tub was coated with paraffin, but this was expensive and hard to do. Depending on the guide, the butter is packed right into the tub or the tub is first line with parchment paper and then packed tightly to the top or close to the top. Then a paper or cloth circle is laid on top and a layer of salt or salt brine water is put on and the cover attached. The salt and brine helped prevent the butter from going bad. Tubs were stored in a cool, dry cellar until they were shipped or used.

TOOLS & MACHINERY - Cream Separators

There are about four different ways to separate cream from milk. At least three of these methods were used in the East Meredith area throughout the 1800s and into the early 1900s. It is unlikely that the Hydraulic Gravity Method would have been used in this area.

Gravity Method - Shallow Pan

This is the easiest, but most time consuming cream separation method. Here are the steps involved:

1. After each milking, pour the milk into shallow pans. Let the pans stand in a cool, clean place for at least 36 hours.
2. After about a day and a half, most of the cream will have risen to the surface. Remove cream with a cream skimmer.
3. Store cream in a cool place until enough is collected for churning.

Pluses
~ Inexpensive method.

Negatives
~ A higher percentage of butterfat remains in the milk.
~ Easiest method for cream to be contaminated.

Gravity Method - Deep Can

This method became popular in the late 1800s. It is also known as the deepset, shotgun, Swedish or Cooley system.

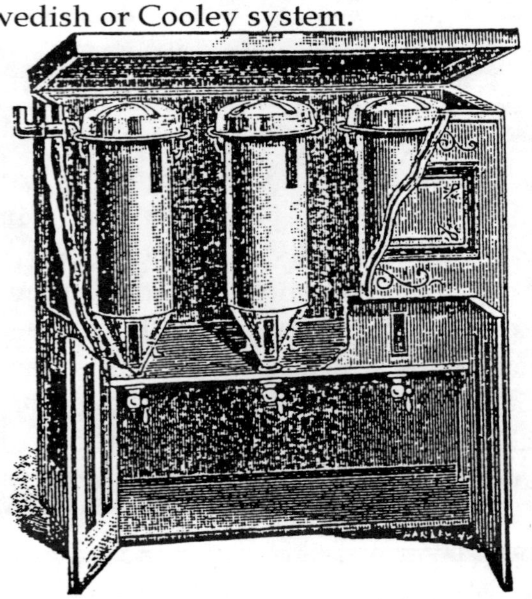

1. Fill cooler or vat with cold water, 40°F is the best temperature. Ice can be added as needed.
2. Pour milk into cans, cover tightly and lower into cooler. Leave for at least 24 hours.
3. After a day the cream will have risen. Use a tube to siphon off the cream.
4. Store cream in a cool place until enough is collected for churning.

Pluses
~ All but .2% of the butterfat can be removed from the milk.

Negatives
~ Expensive equipment required.
~ Needs access to running cold water and/or ice.

Gravity Method - Hydraulic

This method is used if the skimmed milk is not going to be sold or used for other purposes. If this is not the case, then using this method is unethical as it waters down the milk.

1. Pour milk, warm from the cow, directly into a crock or can.
2. Add cold water, equal to about a quarter or half of the amount of milk.
3. Let stand several hours (about 3 hours) until cream rises.
4. Either siphon off cream, or if the crock has a spigot at the bottom, drain off watery skim milk.

Pluses
~ Fast and inexpensive method.

Negatives
~ Watery milk cannot be used afterward.
~ Cream tends to be watery and hard to churn.

Centrifugal Separator Method

This method was almost universally used in East Meredith area dairy farms after the turn-of-the-century. The machine was first developed in Europe in the late 1870s, and the rights in the United States were bought by the DeLaval Company which is still in existence today. The machine uses the centrifugal force of the rotating parts to separate the cream from the milk.

1. Start to turn machine slowly, gradually increasing the speed to 40 to 60 revolutions per minute (many separators have a warning bell when cranked too slow or fast). Maintain speed.
2. Pour some warm water into the regulator cover to warm machine. Water will exit the skim milk spout.
3. Pour milk, warm from the cow, in supply can with faucet closed.
4. Once half of the water has passed through the machine, turn on the milk faucet while maintaining speed. Add milk to supply can as needed until all is separated.
5. Once all the milk has passed through, pour in two quarts of warm water to wash out any left over butterfat.
6. Let go of crank and let the machine slow down on its own. This will reduce wear on moving parts caused by forcing the machine to stop.

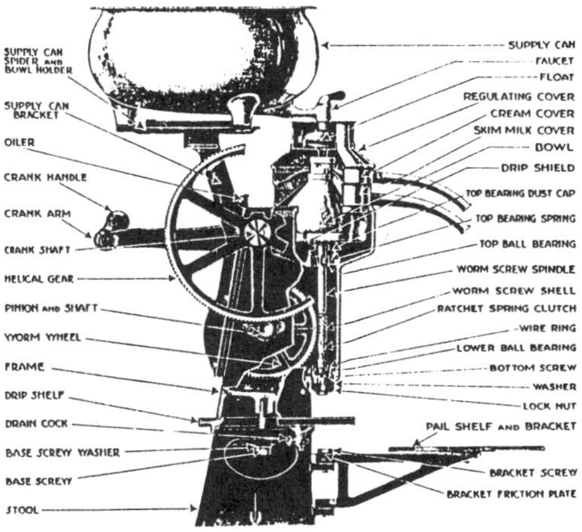

Pluses
~ Produces cream with the highest possible butterfat.
~ Separates cream rapidly.

Negatives
~ Expensive machine.
~ Requires extensive cleaning.

Tools & Machinery - Butter Churns

Butter churns come in almost every shape, size and configuration imaginable. Every new design claimed it was better than the last, and it is all but impossible to say which churn was best. It is important to note, though, that glass or crockery churns are preferred today, as they are easier to clean and keep clean. These first two were the earliest ways of making butter:

Container Agitation:

This is probably the first style of churn ever invented. Cream was placed in a skin bag. With the bag held closed, it was shaken until the butter formed. Later, a jar was used in the same way. Originally the jar was made of clay or crockery with a stopper in its mouth, but today any old glass jar with lid can be used.

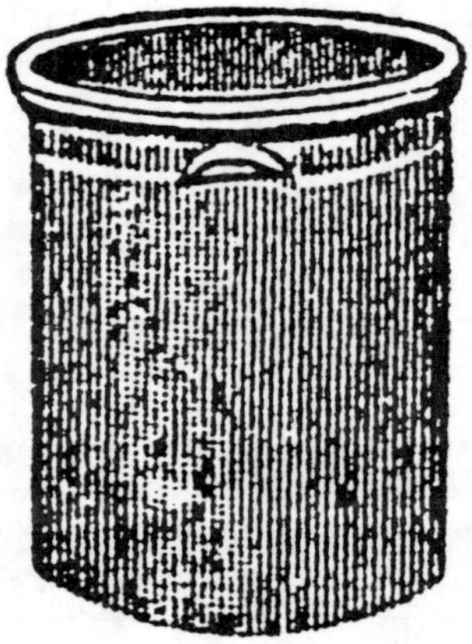

Plunger or Dasher Churn:

This churn type has been used since about 1400 A.D. Traditionally, the churn was made of wood by a cooper. A plunger handle with a dasher at the bottom was pushed up and down in the churn to form the butter. Later crockery and even tin churns were made on this design.

By the mid- and late-1800s, people began to develop other types of churns. The examples below and on the next page cover some of the major and not-so-major designs:

Barrel Churn:

This churn is literally made out of a barrel set upon a framework with an axle at is middle. The milk was put inside and the handle cranked. The barrel rotated end-over-end (though some earlier one rotated along their length). They often had some sort of slatwork or paddles inside to help churn the butter

Box Churns:

There are two types of box churns:

1. This was a box with a four armed set of rotating paddles inside. As the outside crank was turned, they agitated the cream inside. This was also known as a tub churn.
2. This type worked basically like a barrel churn, except the churn was shaped like a box instead of a barrel.

Union Churn:

At the turn-of-the-century, Sears, Roebuck and Company (and other companies) sold this type of churn. They advertised that it would churn, wash, work and salt your butter without removing it from the machine. The claim was probably exaggerated since this type of churn is not often found anymore.

Cylinder Churn:

This churn is similar to the tub type of box churn, but its round shape makes it more efficient. Inside is a rotating double dasher to agitate the cream

Cradle or Swing Churn:

This churn was made of a box, usually with rounded ends, suspended from a framework by rods, chains or rope. By this means the churn was able to rock back and forth like a cradle. These churns often had no paddles or slatwork since the action of the cream being rocked was enough to form butter. It was also easier to clean.

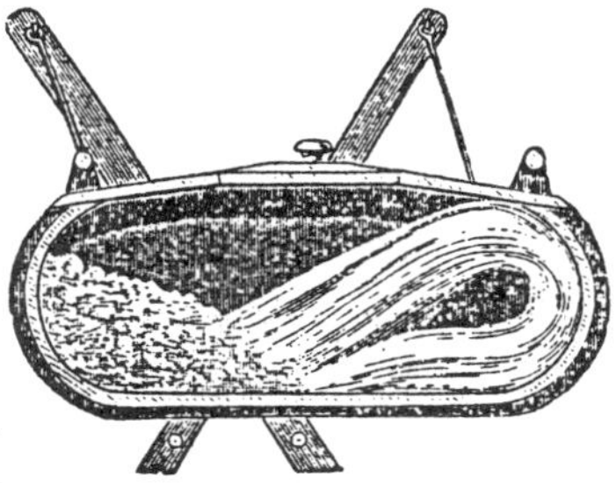

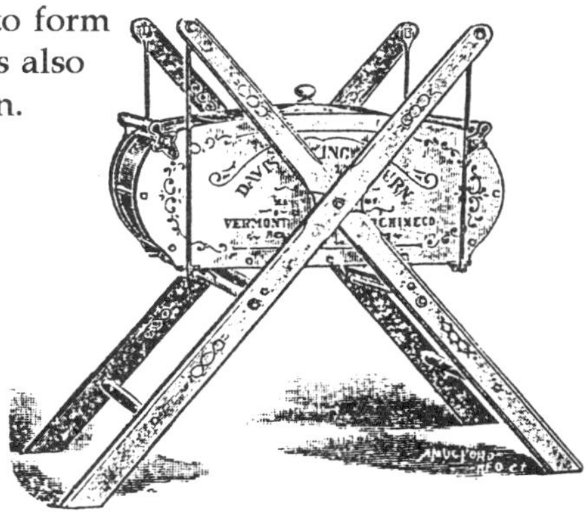

Rotary Dasher Churn:

In this churn, vertical slats are set to force the cream to the center of the churn. A type like this was often used in the 1920s to the 1940s by farmwives who were making a little butter for home use. Many were made with jars large enough to hold only a gallon of cream. It had gears and a crank on the lid to power the dasher.

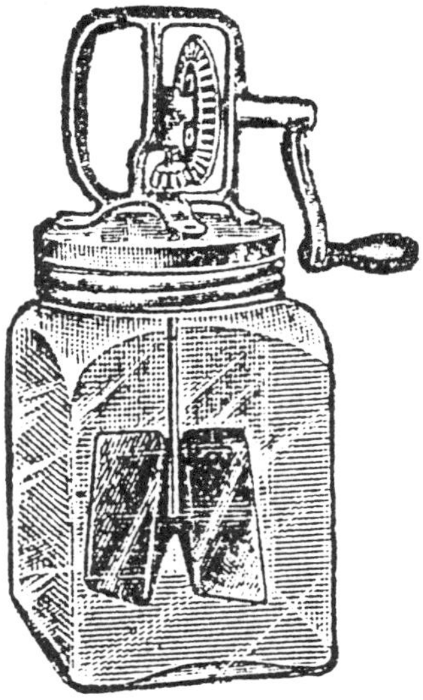

Tools & Machinery - Miscellaneous Equipment

Besides cream separating equipment and churning machines, there were many other tools that could be used when making butter. Here are just a few:

Milking

It should not be forgotten that to get the cream for churning into butter, a cow had to be milked first. There are two major ways this has been accomplished in the past and today.

Milk Pail

The original way to milk a cow was by hand, so of course, some sort of container was needed to catch the milk. Originally, a wooden pail was used. Farmers found it hard to keep these pails clean, so when metal pails became available they became the material of choice. Metal pails were lighter and easier to clean. Some pails came with covers to help keep out dirt and hair, while some had a built in milk strainer which was used when pouring milk into storage containers. The strainer was necessary for cleaning out any dirt and hair that did make it into the milk.

Milking Machine

People were experimenting with the idea of a milking machine by at least the mid-1800s, maybe even earlier. Early machines created suction or pressure using hand or foot powered vacuum pumps or rollers. It wasn't until the gasoline engine became affordable for farm use, though, that the milk machine was widely accepted and used. Today large dairy farms use milking machines almost exclusively. The machine pictured here was the type typically used in the East Meredith area in the 1920s.

Babcock Test

The Babcock Test was available by the 1890s. It was developed by Dr. S.M. Babcock from the state of New York. The test used a machine set up with glass vials in which milk or cream was placed along with a special acid. The vials were whirled by a crank. This allowed a farmer to accurately determine the fat content of his milk or cream. Knowing this helped farmers improve their herd by choosing and breeding the best fat producing cattle, and provided an accurate measure on which to base milk and cream prices.

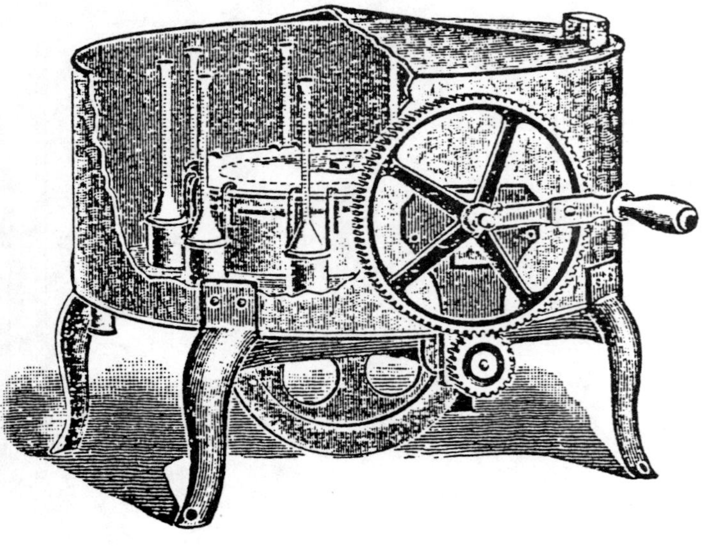

Skimmers, Paddles and Forks

When a gravity method of cream separation was used, especially with the shallow pan method, a perforated metal skimmer was needed to skim the cream from the top of the milk after it had risen.

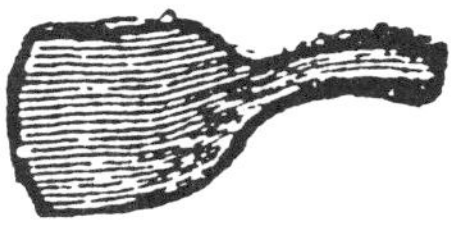

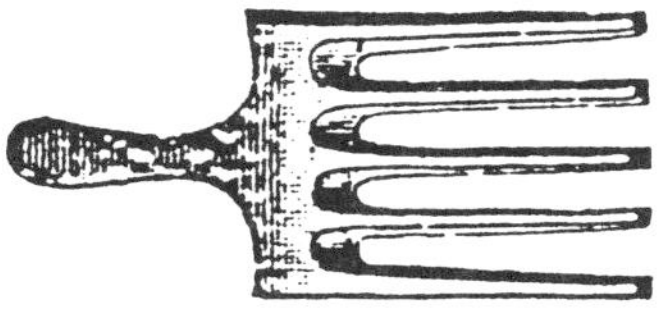

Wooden paddles and forks helped transfer butter from the churn to the working surface, and then from there to the container used for storing or shipping the butter.

Butter Workers

After the butter formed in the churn and was washed, it had to be "worked," literally kneaded like bread. This forced out any excess water and buttermilk. Working butter was traditionally done with the hands in a bowl.

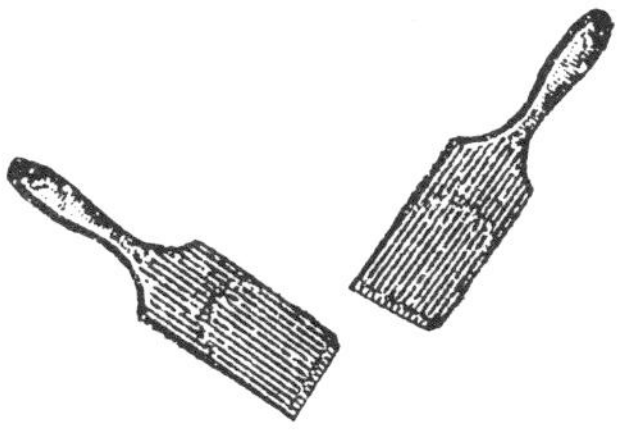

Butter hands or Scotch hands were developed to work butter on a flat surface so the butter wouldn't have to be touched by human hands.

By the late 1800s, butter working was developed even further with the introduction of butter working machines. These usually employed some sort of work surface with a drain and a roller or lever to "work" the butter. Some rollers were smooth while others were fluted or corrugated. They ranged in size from small table top "cottage butter workers" to factory size, machine driven workers.

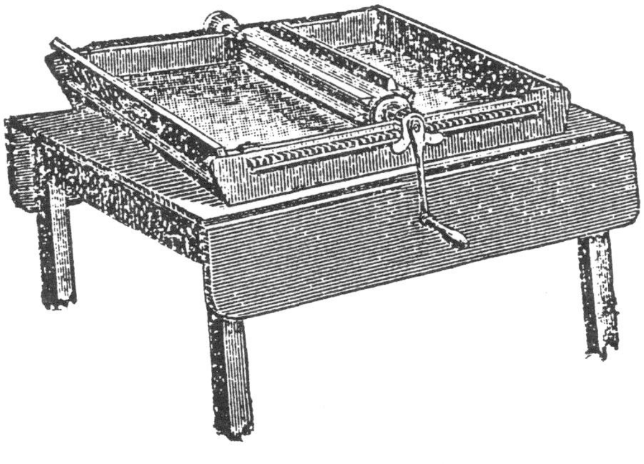

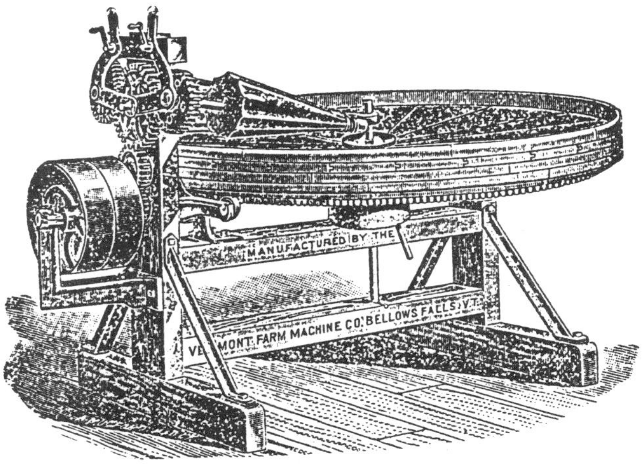

Butter Printers, Molds and Presses

The terms "butter printer" and "butter mold" were interchangeable. Printers or molds were used to form small amounts of butter (between half a pound to two pounds) into a rectangular or round shape. Most butter was pressed into a rectangular block for sale.

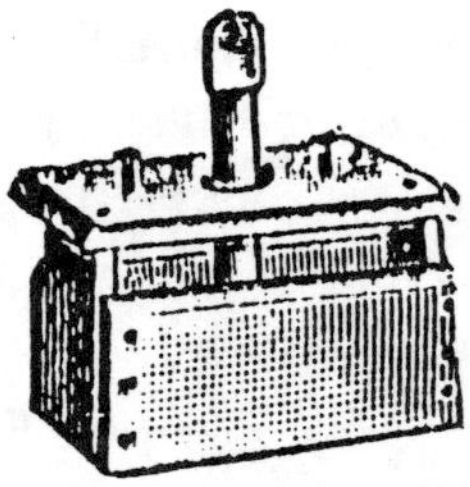

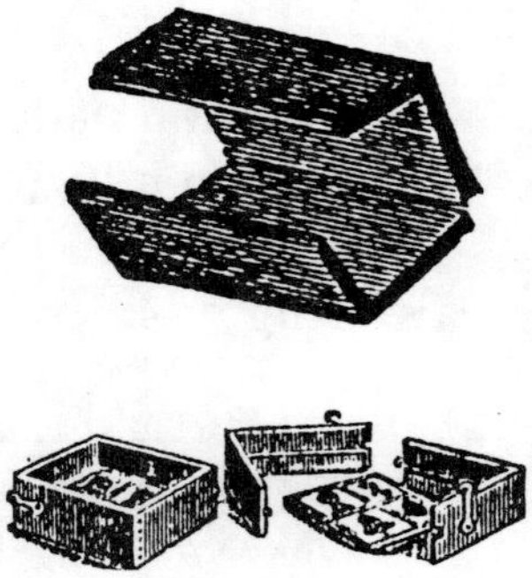

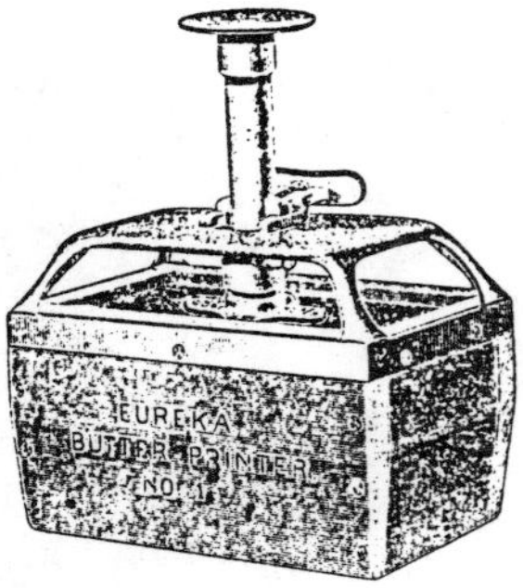

These blocks often had a design stamped into the top in such a way as to indicate where to cut the butter into smaller amounts, for example, where to cut a pound block into quarters or halves. These designs also made the butter more attractive for sale.

Fancy or "gilt edge" butter (meaning the best quality butter) was often molded in round printers or molds.

Butter presses also molded blocks of butter, but a more involved machine was used. A hand or foot powered lever pressed the butter tight into the mold. Like the printers and molds, designs to stamp the top of butter could also be used on a butter press.

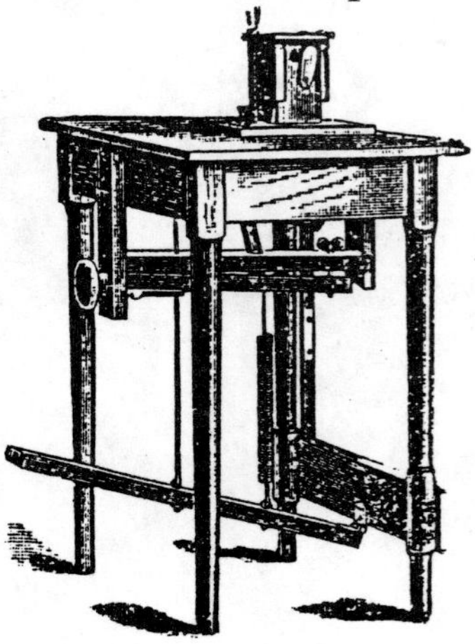

Butter Storage, Shipping and Sale

Butter was not always packed into butter tubs for storage, shipping and sale. There were many types of packaging. These are only a few examples:

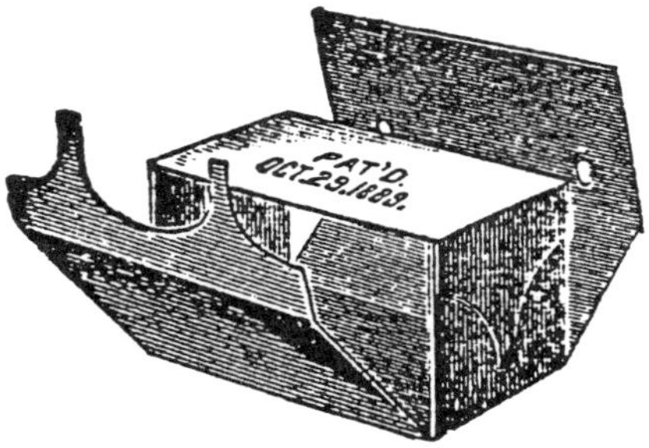

Print Butter Veneer Wrapper
Butter made with a mold was often wrapped in parchment or wax paper, but a pre-cut and scored wrapper made of veneer wood could also be used. This folded to make a package for the butter

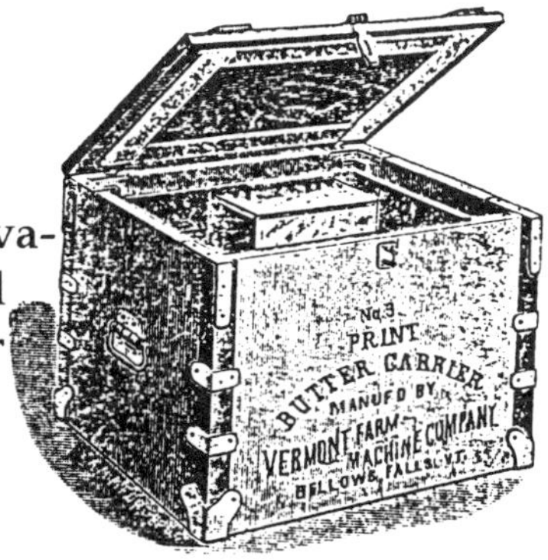

Print Butter Carrier
To carry molded butter and keep it chilled, a wood or galvanized iron box was used. It was often insulated, had dead air spaces and/or a place to put ice to help keep the butter cold.

Round Packages for Butter
These were made of veneer wood, with sizes ranging from a half to ten pounds. The butter was usually taken out of the package by cutting the side seam of the container and removing the veneer.

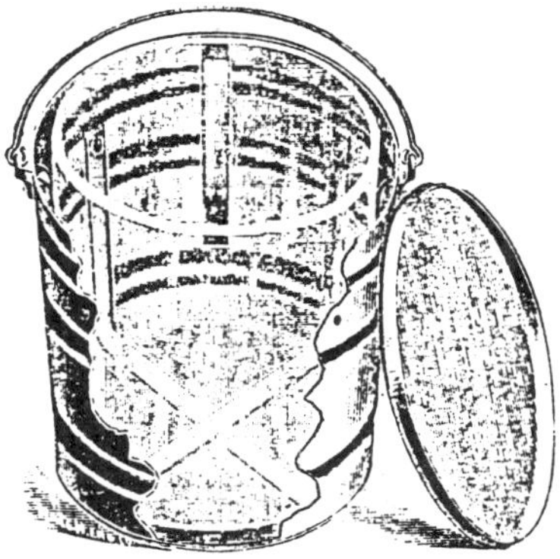

Butter Packages
Besides non-returnable butter tubs, butter was also shipped in butter packages that were shipped back to their point of origin after they were emptied. They were shaped like a butter tub outside (though sometime the wood was in the form of veneer not staves), but inside was a smaller tin can or glass pail. The air space between helped insulate the butter during shipment and the wood outer pail helped protect the inner pail.

Crockery Jars (also known as Stone Jars)
For home use, butter was often packed in crockery or stoneware jars. They were stored in a cool cellar where the crockery would help keep the butter fresh longer.

Tools & Machinery - Power Sources

Besides the obvious power source - human muscle power - there were many other ways to run the machines in a farm creamery (a building were milk was processed on the farm).

Animal Power

Many farmers in the East Meredith area used animal power, usually dogs, to run their separators, churns and butter workers. Local builders like the Flower Brothers and larger companies like the Vermont Farm Machine Company in Bellows Falls, Vermont, made animal treadmills and other types of power sources that could be powered by horses, oxen, sheep or goats. The animal was trained to walk on a treadmill, known as a "power." As the treads moved, a wheel turned which either powered a lever up-and-down or rotated a pulley to provide power for a churn or separator.

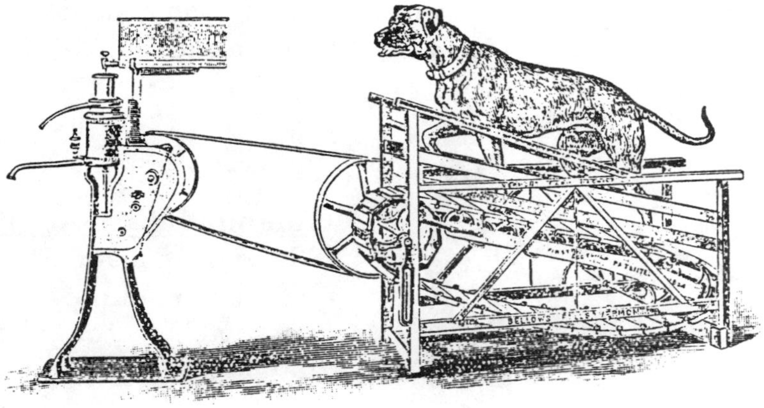

This series of quotes comes from a diary kept by Elizabeth Hanford of East Meredith in 1886:

Saturday, July 10, 1886

We lost our Old Dog Rover, a Pet for us all & our churner. Sick three week.

Thursday, July 15, 1886

[Fred Hagar] & Charlie went to Hope Turners. Got a Dog to Churn.

Saturday, July 17, 1886

Levi took the Dog back to Turner. Did not like to Churn.

Thursday, July 22, 1886

Levi & Brother Will went & Bought a Dog of Mr. Boles.

Water Power

Besides running the mill in East Meredith, D.J. Hanford also had a farm. He made butter in his own creamery. Since D.J. had access to water power, he used that instead of dog power. Any type of water wheel could be used. D.J. Hanford used a turbine set up behind his house in a pit. Some cream separator companies, like the Vermont Farm Machine Company, offered separators that came with a separate pulley attachment so a flat belt could be looped around it and attached to a water wheel or other power source.

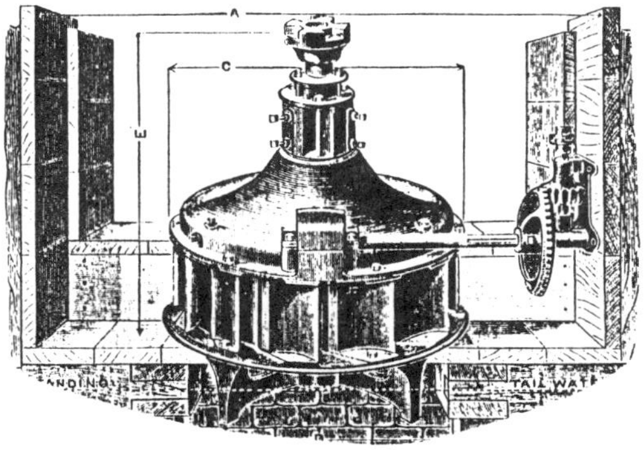

Steam Power

For large dairy farms, a steam engine might be used to run creamery machines. The Vermont Farm Machine Company offered a cream separator with a steam engine. Any machine that could be run using a flat belt attached to a water wheel or animal power, could also be powered by a steam engine. This quote comes from the East Meredith section of the Delaware County Dairyman newspaper:

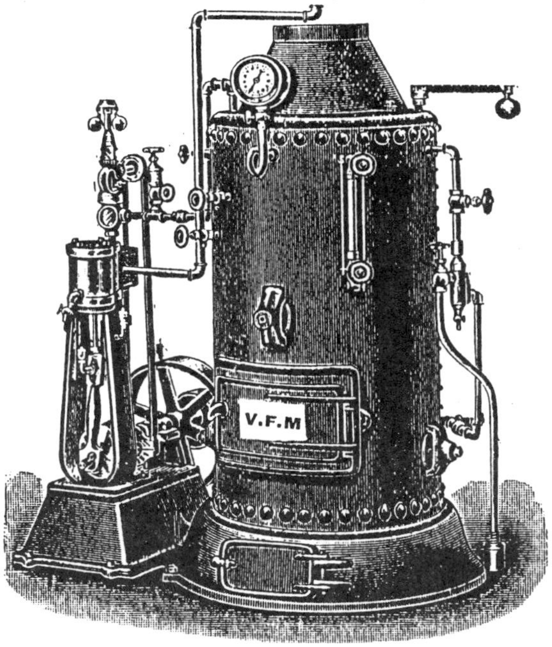

Friday - - March 29, 1895

Homer Wheat of Croton was in town last week placing a large Sharples separator for Mr. James Hamilton of this place. The separator is one of 600 pounds capacity. Mr. Hamilton has one of the finest dairy farms in this section, and a fine herd. An engine will furnish power. A power butter worker will be added. This is the age of improvements; the old, slow methods must go to the wall.

Gasoline Power

After the turn-of-the-century, gasoline engines became available at an affordable price to farmers. These engines helped power many different farm machines, including dairy machines.

The Yearly Calendar - First Cow to Freshen

When a cow "freshens," she gives birth to a calf and her milk begins to flow. Today, farmers plan their breeding schedule so they are milking at least part of their herd year round. In the past, due to feeding practices, markets and breeds, farmers bred their cattle to freshen in the early spring. The first cow to freshen marked the beginning of the dairy season.

Almost every year, Elizabeth Hanford noted in her diary when her first cow freshened. She also wrote down when the last cow freshened so that all their cows could be milked. Sometimes there was quite a difference between the days the first and last cow freshened (as recorded below). Elizabeth did not always keep consistent dairy records, so some years are missing below for lack of information. Black circles mark the first and last cow to freshen.

	1879	1881	1883	1884	1885	1886
January	1 2 3 4 5 6 7 8 9 10 11 12 13 14 15 16 17 18 19 20 21 22 23 24 25 26 27 28 29 30 31	1 2 3 4 5 6 7 8 9 10 11 12 13 14 15 16 17 18 19 20 21 22 23 24 25 26 27 28 29 30 31	1 2 3 4 5 6 7 8 9 10 11 12 13 14 15 16 17 18 19 20 21 22 23 24 25 26 27 28 29 30 31	1 2 3 4 5 6 7 8 9 10 11 12 13 14 15 16 17 18 19 20 21 22 23 24 ● 26 27 28 29 30 31	1 2 3 4 5 6 7 8 9 10 11 12 13 14 15 16 17 18 19 20 21 22 23 24 25 26 27 28 29 30 31	1 2 3 4 5 6 7 8 9 10 11 12 13 14 15 16 17 18 19 20 21 22 23 24 25 26 27 28 29 30 31
February	1 2 3 4 5 6 7 8 9 10 11 12 13 14 15 16 17 18 19 20 21 22 23 24 25 26 27 ●	1 2 3 4 5 6 7 8 9 10 11 12 13 14 15 16 17 18 19 20 21 22 23 24 25 26 27 28	1 2 3 4 5 6 7 8 9 10 11 12 13 ● 15 16 17 18 19 20 21 22 23 24 25 26 27 28	1 2 3 4 5 6 7 8 9 10 11 12 13 14 15 16 17 18 19 20 21 22 23 24 25 26 27 28 29	1 2 3 4 5 6 7 ● 9 10 11 12 13 14 15 16 17 18 19 20 21 22 23 24 25 26 27 28	1 2 3 ● 5 6 7 8 9 10 11 12 13 14 15 16 17 18 19 20 21 22 23 24 25 26 27 28
March	1 2 3 4 5 6 7 8 9 10 11 12 13 14 15 16 17 18 19 20 21 22 23 24 25 26 27 28 29 30 31	1 ● 3 4 5 6 7 8 9 10 11 12 13 14 15 16 17 18 19 20 21 22 23 24 25 26 27 28 29 30 31	1 2 3 4 5 6 7 8 9 10 11 12 13 14 15 16 17 18 19 20 21 22 23 24 25 26 27 28 29 30 31	1 2 3 4 5 6 7 8 9 10 11 12 13 14 15 16 17 18 19 20 21 22 23 24 25 26 27 28 29 30 31	1 2 3 4 5 6 7 8 9 10 11 12 13 14 15 16 17 18 19 20 21 22 23 24 25 26 27 28 29 30 31	1 2 3 4 5 ● 7 8 9 10 11 12 13 14 15 16 17 18 19 20 21 22 23 24 25 26 27 28 29 30 31
April	1 2 3 4 5 6 7 8 9 10 11 12 13 14 15 16 17 18 19 20 21 22 23 24 25 26 27 28 29 30	1 2 3 4 5 6 7 8 9 10 11 12 ● 14 15 16 17 18 19 20 21 22 23 24 25 26 27 28 29 30	1 2 3 4 5 6 7 ● 9 10 11 12 13 14 15 16 17 18 19 20 21 22 23 24 25 26 27 28 29 30	1 2 3 4 5 6 7 8 9 10 11 12 13 14 15 ● 17 18 19 20 21 22 23 24 25 26 27 28 29 30	1 2 3 4 5 6 7 8 9 10 11 12 13 14 15 16 17 18 19 20 ● 22 23 24 25 26 27 28 29 30	1 2 3 4 5 6 7 8 9 10 11 12 13 14 15 16 17 18 19 20 21 22 23 24 25 26 27 28 29 30
May	1 2 3 4 5 6 7 8 9 10 11 12 13 14 15 16 ● 18 19 20 21 22 23 24 25 26 27 28 29 30 31	1 2 3 4 5 6 7 8 9 10 11 12 13 14 15 16 17 18 19 20 21 22 23 24 25 26 27 28 29 30 31	1 2 3 4 5 6 7 8 9 10 11 12 13 14 15 16 17 18 19 20 21 22 23 24 25 26 27 28 29 30 31	1 2 3 4 5 6 7 8 9 10 11 12 13 14 15 16 17 18 19 20 21 22 23 24 25 26 27 28 29 30 31	1 2 3 4 5 6 7 8 9 10 11 12 13 14 15 16 17 18 19 20 21 22 23 24 25 26 27 28 29 30 31	1 2 3 4 5 6 7 8 9 10 11 12 13 14 15 16 17 18 19 20 21 22 23 24 25 26 27 28 29 30 31
	10 cows	11 Cows	10 Cows	10 Cows	10 Cows	10 Cows

THE YEARLY CALENDAR - Churning Butter

The cream from one cow did not necessarily give enough butter to pack a butter tub. Elizabeth Hanford's diaries indicate that she did not churn every day. It appears that she waited to collect a certain amount and then she churned. It is interesting to watch her diary accounts, for she often noted the first time she churned in a year, and sometimes the last time she churned. Below, black circles denote dates that Elizabeth recorded specifically as a first or last churning. Other dates under shaded circles are the first or last date that she noted churning, though she did not write it was the first or last. Compare these dates with the previous page, though some years are missing due to lack of information.

1879	1880	1881	1882	1883	1884
January 1 2 3 4 5 6 7 8 9 10 11 12 13 14 15 16 17 18 19 20 21 22 23 24 25 26 27 28 29 30 31	January 1 2 3 4 5 6 7 8 9 10 11 12 13 14 15 16 17 18 19 20 21 22 23 24 25 26 27 28 29 30 31	January 1 2 3 4 5 6 7 8 9 10 11 12 13 14 15 16 17 18 19 20 21 22 23 24 25 26 27 28 29 30 31	January 1 2 3 4 5 6 7 8 9 10 11 12 13 14 15 16 17 18 19 20 21 22 23 24 25 26 27 28 29 30 31	January 1 2 3 4 5 6 7 8 9 10 11 12 13 14 15 16 17 18 19 20 21 22 23 24 25 26 27 28 29 30 31	January 1 2 3 4 5 6 7 8 9 10 11 12 13 14 15 16 17 18 19 20 21 22 23 24 25 26 27 28 29 30 31
February 1 2 3 4 5 6 7 8 9 10 11 12 13 14 15 16 17 18 19 20 21 22 23 24 25 26 27 28	February 1 2 3 4 5 6 7 8 9 10 11 12 13 14 15 16 17 18 19 20 21 22 23 24 25 26 27 28 29	February 1 2 3 4 5 6 7 8 9 10 11 12 13 14 15 16 17 18 19 20 21 22 23 24 25 26 27 28	February 1 2 3 4 5 6 7 8 9 10 11 12 13 14 15 16 17 ● 19 20 21 22 23 24 25 26 27 28	February 1 2 3 4 5 6 7 8 9 10 11 12 13 14 15 16 17 18 19 ● 21 22 23 24 25 26 27 28	February 1 2 3 4 5 6 7 8 9 10 11 12 13 14 15 16 17 18 19 20 21 22 23 24 25 26 27 28 29
March 1 2 3 4 ● 6 7 8 9 10 11 12 13 14 15 16 17 18 19 20 21 22 23 24 25 26 27 28 29 30 31	March 1 2 3 4 5 6 7 8 9 10 11 12 13 14 15 16 17 18 19 20 21 22 23 24 25 26 27 28 29 30 31	March 1 2 3 4 5 6 7 ● 9 10 11 12 13 14 15 16 17 18 19 20 21 22 23 24 25 26 27 28 29 30 31	March 1 2 3 4 5 6 7 8 9 10 11 12 13 14 15 16 17 18 19 20 21 22 23 24 25 26 27 28 29 30 31	March 1 2 3 4 5 6 7 8 9 10 11 12 13 14 15 16 17 18 19 20 21 22 23 24 25 26 27 28 29 30 31	March 1 2 3 4 5 6 7 8 9 10 11 12 13 14 15 16 17 18 19 20 21 22 23 24 25 26 27 28 29 30 31
November 1 2 3 4 5 6 7 8 9 10 11 12 13 14 15 16 17 18 19 20 21 22 23 24 25 26 27 28 29 30	November 1 2 3 4 5 6 7 8 9 10 11 12 13 14 15 16 17 18 19 20 21 22 23 24 25 26 27 28 29 30	November 1 2 3 4 5 6 7 8 9 10 11 12 13 14 15 16 17 18 19 20 21 22 23 24 25 26 27 28 29 30	November 1 2 3 4 5 6 7 8 9 10 11 12 13 14 15 16 17 18 19 20 21 22 23 24 25 26 27 28 29 30	November 1 2 3 4 5 6 7 8 9 10 11 12 13 14 15 16 17 18 19 20 21 22 23 24 25 26 27 28 29 30	November 1 2 3 4 5 6 7 8 9 10 11 12 13 14 15 16 17 18 19 20 21 22 23 24 25 26 27 28 29 30
December 1 2 3 4 5 6 7 8 9 10 11 12 13 14 15 16 17 18 19 20 21 22 23 24 25 26 27 28 29 30 31	December 1 2 3 4 5 6 7 8 9 10 11 12 13 14 15 16 17 18 19 20 21 22 23 24 25 26 27 28 29 30 31	December 1 2 3 4 5 6 7 8 9 10 11 12 13 14 15 16 17 18 19 20 21 22 23 24 25 26 27 28 29 30 31	December 1 2 3 4 5 6 7 8 9 10 11 12 13 14 15 16 17 18 19 20 21 22 23 24 25 26 27 28 29 30 31	December 1 2 3 4 5 6 7 8 9 10 11 12 13 14 15 16 17 18 19 20 21 22 23 24 25 26 27 28 29 30 31	December 1 2 3 4 5 6 7 8 9 10 11 12 13 14 15 16 17 18 19 20 21 ● 23 24 25 26 27 28 29 30 31

Churning had to be done when you had enough cream to make butter, but Elizabeth may have had a scheduled churning day - a day on which she would try to make butter if she could. Of all the churning days Elizabeth Hanford noted in her diary (and she probably did not record them all), Monday was her day of choice. Even though that was her wash day, 58% of the churning days that were written in her diary were also on Monday. Tuesday was a distant second, being used 15% of the time for churning. But cows gave milk every day, often in varying amounts, so when enough cream was collected it was churned. There are times, at the peak of the dairy season, where Elizabeth churned twice in a day. Elizabeth even recorded that she churned on a Sunday once! Sunday was still a day of rest and religious observance - the Hanfords even let their maple sap buckets run over if the sap ran too fast on Sunday. Since Elizabeth avoided churning on Sunday, it makes sense that her churning day was Monday.

THE YEARLY CALENDAR - Packing Tubs & Firkins

Elizabeth Hanford packed 50 pound butter tubs in the spring and fall. In the spring, cows were freshening at different times, and in the fall cows were going dry. This meant that in the spring and fall, she didn't always have enough butter to pack 100 pound firkins fast enough. She saved firkins for packing in the summer months. It was a pattern used since at least the mid-1800s.

Elizabeth was also commissioned to pack butter for certain friends and relatives, probably for their own personal use through the winter. This she did in the fall, packing their butter into butter tubs and crockery jars. She made butter for the Evans family in East Meredith, for George Flower in White Plains near New York City (the brother of her daughter-in-law), and P.R. Young, a well-known photographer in Oneonta, NY (and according to her diaries, a long time family friend). She also occasionally made butter for at least one of the local general stores, J.H. Thompson's on the corner in East Meredith.

It is difficult to follow each tub and firkin as Elizabeth Hanford filled them. Sometimes she noted when she started a container of butter and then when she finished filling it, but this was not often the case. From the information she *has* left in her diaries, we can say that it took her anywhere from 5 to 22 days to fill a butter tub. Her average time for filling a tub was about 9 days. The longer periods of time between commencing a tub, as she called it, to filling it were in November and December when cows were drying up. To fill a firkin, it took her from 9 to 16 days, but she averaged 12 days.

MYSTERY:

Why did Elizabeth Hanford switch from
packing butter tubs to firkins and then back to butter tubs?

Some have suggested it was because of larger quantities of butter during the summer.
Why not just fill more butter tubs faster?
OR
Does the switch have to do with a change in the quality of butter
between spring and fall butter compared to summer butter?

The calendars on the next page show when Elizabeth Hanford commenced her first butter tubs and firkins and when she filled the last each year.

The black circles denote the spring and fall butter tubs, first and last, while the shaded circles show the first and last firkins. Sometimes there is quite a lag between last firkin and first fall butter tub; this is due to the fact that Elizabeth often noted when she started her last firkin, but not when she finished it. To save space, certain months have been left out. In those months Elizabeth neither began nor finished production of butter, though she was continuing the production. Compare these dates with the previous pages to get the full picture of Elizabeth Hanford's yearly dairy calendar.

Month	1879	1880	1881	1882	1883	1884
February	1 2 3 4 5 6 7 8 9 10 11 12 13 14 15 16 17 18 19 20 21 22 23 24 25 26 27 28	1 2 3 4 5 6 7 8 9 10 11 12 13 14 15 16 17 18 19 20 21 22 23 24 25 26 27 28 29	1 2 3 4 5 6 7 8 9 10 11 12 13 14 15 16 17 18 19 20 21 22 23 24 25 26 27 28	1 2 3 4 5 6 7 8 9 10 11 12 13 14 15 16 17 18 19 20 21 22 ● 24 25 26 27 28	1 2 3 4 5 6 7 8 9 10 11 12 13 14 15 16 17 18 19 20 21 22 23 24 25 ● 27 28	1 2 3 4 5 6 7 8 9 10 ● 12 13 14 15 16 17 18 19 20 21 22 23 24 25 26 27 28 29
March	1 2 3 4 5 6 7 8 9 10 ● 12 13 14 15 16 17 18 19 20 21 22 23 24 25 26 27 28 29 30 31	1 2 3 4 5 6 7 8 9 10 11 12 13 14 ● 16 17 18 19 20 21 22 23 24 25 26 27 28 29 30 31	1 2 3 4 5 6 7 8 9 10 11 12 13 14 15 ● 17 18 19 20 21 22 23 24 25 26 27 28 29 30 31	1 2 3 4 5 6 7 8 9 10 11 12 13 14 15 16 17 18 19 20 21 22 23 24 25 26 27 28 29 30 31	1 2 3 4 5 6 7 8 9 10 11 12 13 14 15 16 17 18 19 20 21 22 23 24 25 26 27 28 29 30 31	1 2 3 4 5 6 7 8 9 10 11 12 13 14 15 16 17 18 19 20 21 22 23 24 25 26 27 28 29 30 31
May	1 2 3 4 5 6 7 8 9 10 11 12 13 14 15 16 17 18 19 20 ● ◍ 23 24 25 26 27 28 29 30 31	1 2 3 4 5 6 7 8 9 10 11 12 13 14 15 16 17 18 19 20 21 ● 23 24 ◍ 26 27 28 29 30 31	1 2 3 4 5 6 7 8 9 10 11 12 13 14 15 16 17 18 19 20 21 22 23 24 25 ● ◍ 28 29 30 31	1 2 3 4 5 6 7 8 9 10 11 12 13 14 15 16 17 18 19 20 21 22 23 24 25 26 27 28 29 30 31	1 2 3 4 5 6 7 8 9 10 11 12 13 14 15 16 17 18 19 20 21 22 23 24 25 ● 27 ◍ 29 30 31	1 2 3 4 5 6 7 8 9 10 11 12 13 14 15 16 17 18 19 20 21 22 23 24 25 26 27 ● 29 ◍ 31
June	1 2 3 4 5 6 7 8 9 10 11 12 13 14 15 16 17 18 19 20 21 22 23 24 25 26 27 28 29 30	1 2 3 4 5 6 7 8 9 10 11 12 13 14 15 16 17 18 19 20 21 22 23 24 25 26 27 28 29 30	1 2 3 4 5 6 7 8 9 10 11 12 13 14 15 16 17 18 19 20 21 22 23 24 25 26 27 28 29 30	1 ● ◍ 4 5 6 7 8 9 10 11 12 13 14 15 16 17 18 19 20 21 22 23 24 25 26 27 28 29 30	1 2 3 4 5 6 7 8 9 10 11 12 13 14 15 16 17 18 19 20 21 22 23 24 25 26 27 28 29 30	1 2 3 4 5 6 7 8 9 10 11 12 13 14 15 16 17 18 19 20 21 22 23 24 25 26 27 28 29 30
July	1 2 3 4 5 6 7 8 9 10 11 12 13 14 15 16 17 18 19 20 21 22 23 24 ◍ 26 27 28 29 30 31	1 2 3 4 5 6 7 8 9 10 11 12 13 14 15 16 ◍ 18 19 20 21 22 23 24 25 26 27 28 29 30 31	1 2 3 4 5 6 7 8 9 10 11 12 13 14 15 16 17 18 19 20 21 22 23 24 25 26 27 28 29 30 31	1 2 3 4 5 6 7 8 9 10 11 12 13 14 15 16 17 18 19 20 21 22 23 24 25 26 27 28 29 30 31	1 2 3 4 5 6 7 8 9 10 11 12 13 14 15 16 ◍ 18 19 20 21 22 23 24 25 26 27 28 29 30 31	1 2 3 4 5 6 7 8 9 10 11 12 13 14 15 16 17 18 19 20 21 22 23 24 ◍ 26 27 28 29 30 31
August	1 2 3 4 ● 6 7 8 9 10 11 12 13 14 15 16 17 18 19 20 21 22 23 24 25 26 27 28 29 30 31	1 2 ● 4 5 6 7 8 9 10 11 12 13 14 15 16 17 18 19 20 21 22 23 24 25 26 27 28 29 30 31	1 2 3 4 5 6 7 ◍ ● 10 11 12 13 14 15 16 17 18 19 20 21 22 23 24 25 26 27 28 29 30 31	◍ 2 3 4 5 6 7 8 9 10 11 12 13 14 15 16 ● 18 19 20 21 22 23 24 25 26 27 28 29 30 31	1 2 3 4 5 6 7 8 9 10 11 12 13 14 15 ● 17 18 19 20 21 22 23 24 25 26 27 28 29 30 31	1 2 3 4 5 6 7 8 ● 10 11 12 13 14 15 16 17 18 19 20 21 22 23 24 25 26 27 28 29 30 31
November	1 2 3 4 5 6 7 8 9 10 11 12 13 14 15 16 17 18 19 20 21 22 23 24 25 26 ● 28 29 30	1 2 3 ● 5 6 7 8 9 10 11 12 13 14 15 16 17 18 19 20 21 22 23 24 25 26 27 28 29 30	1 2 3 4 5 6 7 8 9 10 11 12 13 14 15 16 17 18 19 20 21 22 23 24 25 26 27 28 29 30	1 2 3 4 5 6 7 8 9 10 11 12 13 14 15 16 17 18 19 20 21 22 23 24 25 26 27 28 29 30	1 2 3 4 5 6 7 8 9 ● 11 12 13 14 15 16 17 18 19 20 21 22 23 24 25 26 27 28 29 30	1 2 3 4 5 6 7 8 9 10 11 12 13 14 15 16 17 18 19 20 21 22 23 24 25 26 27 28 29 30
December	1 2 3 4 5 6 7 8 9 10 11 12 13 14 15 16 17 18 19 20 21 22 23 24 25 26 27 28 29 30 31	1 2 3 4 5 6 7 8 9 10 11 12 13 14 15 16 17 18 19 20 21 22 23 24 25 26 27 28 29 30 31	1 2 ● 4 5 6 7 8 9 10 11 12 13 14 15 16 17 18 19 20 21 22 23 24 25 26 27 28 29 30 31	1 2 3 4 5 6 7 8 9 10 ● 12 13 14 15 16 17 18 19 20 21 22 23 24 25 26 27 28 29 30 31	1 2 3 4 5 6 7 8 9 10 11 12 13 14 15 16 17 18 19 20 21 22 23 24 25 26 27 28 29 30 31	1 ● 3 4 5 6 7 8 9 10 11 12 13 14 15 16 17 18 19 20 21 22 23 24 25 26 27 28 29 30 31

BUTTER PRODUCTION NUMBERS

In a history of Delaware County, New York that was printed in the 1890s, the author wrote that early cows produced about 100 pounds of butter a year, while "modern" cows could produce closer to 400 pounds. As stated earlier, these improvements were due to better breeding and feeding of the farmer's cattle.

Elizabeth Hanford's cows averaged about 190 pounds of a butter a year in the early 1880s. By 1884 and 1885, however, they were averaging 220 to 250 pounds of butter a year. Unfortunately, Elizabeth does not give us enough information to say whether this was a continuing upward trend or a fluke of high production for just those two years. Elizabeth's diaries give us no information on where the Hanford cows came from , how their cattle were bred, how they fed their cattle or how they treated their animals. Without this information it is difficult to call the increase in butter production a upward trend, but it does fit what was written in the history of Delaware County, New York.

Elizabeth Hanford packed butter tubs in the spring and fall, and then packed firkins in the summer. As the production of butter went up, she never increased the number of firkins. Elizabeth averaged 6 firkins throughout the period between 1879 and 1885 What she did do was reduce the time between when she started packing her first firkin and when she finished the last firkin (see previous section). Elizabeth did pack more butter tubs spring and fall as her cows produced more butter:

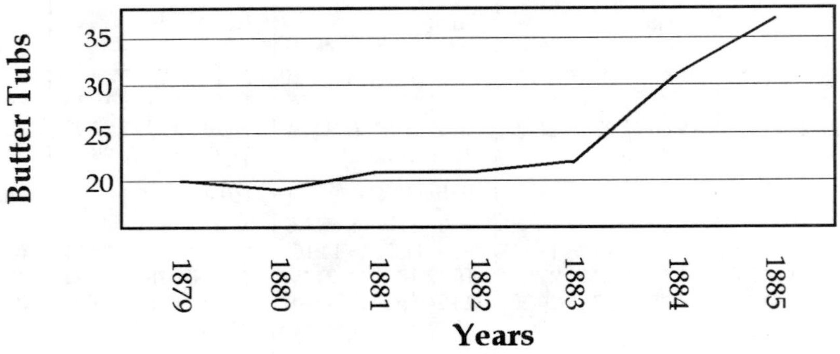

MYSTERY:

When butter production increased,
Elizabeth Hanford increased the amount of butter tubs she filled,
but she didn't increase the number of firkins she filled.
Instead, she shortened the period in which she filled butter tubs.

Why?

PRICES

The prices offered for butter were based on three different variables:

1. The quality of the butter
2. The time of year the butter was sold
3. The freshness or time of year the butter was made

	Extras.	Mediums.	Poor.
New Butter.			
Creameries	a21	15 a19	12 a14
State dairy, tubs and pails	a18	15 a17	12 a14
Imitation creamery	13 a14	10 a12	9 a
Western factory	11½a12	8 a10	7 a7 1-2
Old Butter.			
Creameries, summer make	a14	12 a13	9 a11
Creameries, fall make.	a15	12 a14	10 a11
State dairies, entire	14 a	13 a13	10 a11
Western factory	9 a	8 a	7 a7 1-2

Market Report, New York City, March, 23, 1895

Elizabeth Hanford never mentioned butter quality in her diary, but she was probably fairly consistent in producing a good quality, medium butter (see The Columbian World's Exposition and Butter). The prices she received for her butter varied due to the last two factors listed above, the time of year the butter was sold and its freshness

During the Civil War, Elizabeth Hanford's butter sold for as much as fifty cents a pound, but this was due to war-time inflation. By the 1880s, when we have the best information, the price for Elizabeth's butter ranged from forty-three cents to twelve cents a pound, and averaged a price of twenty-seven cents a pound.

Butter prices were often at their highest during the first few months of the year when most farmers' cows were no longer producing, but prices might often go up in the fall, as well, as cows herds went dry. Elizabeth Hanford had to gage when was the best time to sell her butter. A study of her diaries show that if the prices for butter remained relatively stable throughout the year, Elizabeth might sell all her butter by the end of fall. But if the price of butter dropped dramatically in the fall months, she waited to sell most of her butter until the first few months in the next year.

Elizabeth Hanford's 1882 diary is the only one in which she records exactly how much she made in a year. She wrote:

> Monday, December 11, 1882
>
> *Some snow Squalls We washed & churned not very cold. filled my ninth tub. which makes about sixteen hundred lbs from ten cows in numbr some heifer & sold one early.*

At the very end of the 1882 diary, she wrote:

> *Have made from our ten cows $500.17 doollar*

With 1600 pounds of butter at $500.17 for the year, Elizabeth Hanford averaged thirty-one cents a pound that year. Considering that a worker at Hanford Mills in East Meredith made about a dollar a day at the same time, Elizabeth did very well. Of course, it should also be kept in mind, though, that the men in Elizabeth's family purchased, cared for and fed the cows. Elizabeth did not include these expenses when she said she made over $500.

Using the information that Elizabeth provided in each of her diaries, it appears that she may have averaged about $460 a year, with high years close to $800 and low years nearer to $200.

MYSTERY:

What happened to the money Elizabeth Hanford made with her butter?

DID SHE USE IT FOR HOUSEHOLD EXPENSES?

OR

WAS THE MONEY PUT BACK INTO RUNNING THE FARM, TO PROVIDE FOOD AND SHELTER FOR THEIR HERD?

The Commission Merchant

When Elizabeth Hanford wrote in her 1882 diary that she had earned $500.17 with ten cows, that did not necessarily mean that she had $500.17 in hand. The best markets for selling butter were in the large cities. Rather than travel to these markets to sell butter, most area farmers sent their product to a "commission merchant."

Commission merchants advertised in the local newspapers. They were located in cities like Albany and New York City in New York, and Scranton in Pennsylvania. They offered to sell items (such as butter, cheese, eggs, poultry and other products) for the farmers who sent them. Many advertised that one of their "specialties" was "Delaware County butter."

The reason why Elizabeth Hanford did not receive all the money she earned was because the commission merchants, of course, worked for a price. After they quoted a price per pound on the butter they sold for Elizabeth, they took out what some called "freight and charges" for the shipping of the butter on the railroad and a 5% commission fee for selling the butter. Before a railroad reached a point only three miles from East Meredith, the commission merchant also charged for the cost of transporting the butter from the farm to a railroad station. Butter was usually moved by horse and wagon, and the charge appeared on the receipt as "cartage."

The commission merchant promised to get the best price for the butter he sold, saving the farmer the time and expense of trying to sell their own product. Elizabeth, and her husband Levi, did keep an eye on the market though, switching between different commission merchants as needed. Besides the factors mentioned previously, the prices they received probably also depended on the location of the commission merchant and how well he did his job. Between 1880 and 1886, Elizabeth used six different commission merchants. She made most of her changes back and forth in 1885 and 1886. Elizabeth Hanford sent her tubs to Loughran & Egbert, Seymour, Vannata, Mead, Egbert & Case, Demorest and McCabe. D.J. Hanford dealt with two other merchants - David W. Lewis & Co. and William M. Hussey. The November 23, 1894 advertisements from the Delaware County Dairyman newspaper on the right show some of these names, and include even more.

SEYMOUR & NOE,
Commission Merchants,
Specialty---Del. Co. Butter.
176 Chambers Street, New York.

Quick Sales,
Prompt Returns.
Give Us a Trial.

WE BUY BUTTER.
The T. H. Watts Co., Lim.,
WHOLESALE DEALERS IN
BUTTER. CHEESE, EGGS, ETC,
723, 725 West Lackawanna Ave.,
SCRANTON, PA.
All correspondence will receive prompt attention.

C. B. HILL & CO.,
PRODUCE
COMMISSION MERCHANTS,
314 Greenwich Street, New York.
Fine Fresh Made Dairy, Pail and Tub Butter a Specialty.
Shipments of produce thus received will be sold as soon after arrival as the condition of the market will warrant to advantage, and check positively rendered on each Saturday following sale.
REFERENCE:—New York National Exchange Bank

HEALEY & ELLIS,
Commission Merchants.
BUTTER, CHEESE, EGGS, &c.
359 Greenwich Street, New York,
(Corner Harrison Street.)
REFERENCE:
Irving National Bank, New York.
McClelland & Anderson, Bankers, Detroit, Mich.
W. J. Campbell, LaRue, Ohio.

☞Delaware County Butter a Specialty.
EGBERT & CASE,
PRODUCE
Commission Merchants
80 Warren Street, New York.
☞Refer to IRVING NATIONAL BANK.

Established 1841. Incorporated 1893.
GEO. B. DOUGLAS & CO.,
Produce Commission,
286 Greenwich Street New York.
REFERENCE:
New York National Ex. Bank,
New York City.

Jelliffe, Wright & Co.,
POULTRY, BUTTER, EGGS, Etc.,
284 Washington Street, New York.

Branch for Meats: 22, 24 & 26 Grace Avenue, West Wash. Market.	Branch for Live Stock: At the Union Stock Yards, West 60th Street.

ESTABLISHED 1859.
CHAS. P. MECABE & SON,
PRODUCE
Commission Merchants
Butter, Cheese, Eggs, Etc.
291 Washington Street, New York.

Compare these two receipts from two different commission merchants from two different years in almost the same month. The first one is for Elizabeth Hanford's butter sent by her husband, Levi, in May of 1870. Note that since there was no close railroad station, cartage was charged. The second receipt was for butter sent by D.J. Hanford from his farm in April of 1895.

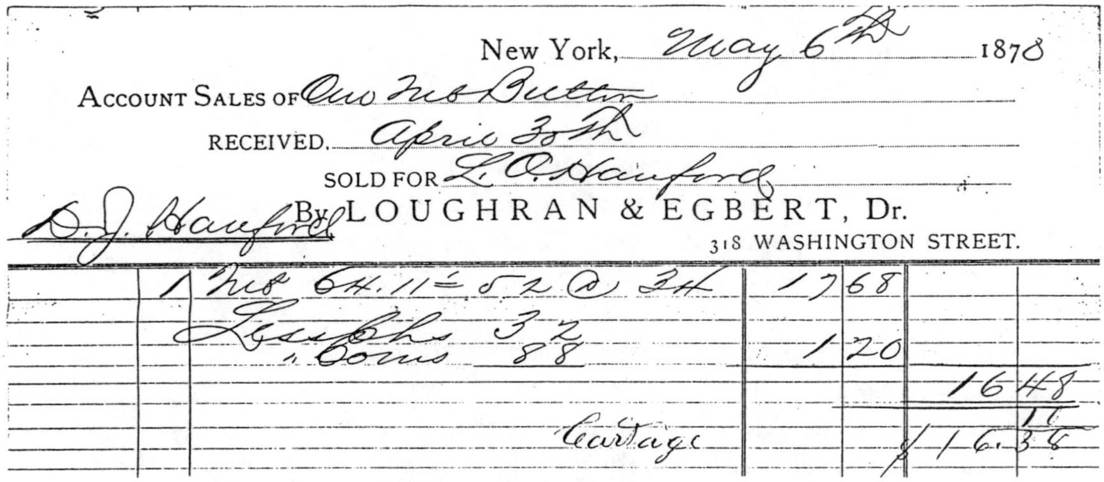

New York, May 6th 1870

ACCOUNT SALES OF One tub Butter

RECEIVED April 30th

SOLD FOR L. O. Hanford

D. J. Hanford By LOUGHRAN & EGBERT, Dr.

318 WASHINGTON STREET.

1 tub 64.11 = 52 @ 34	17 68		
Less chs 32			
Coms 88	1 20		
		16	48
			11
Cartage		$16.	36

It appears as if D.J. Hanford, Levi's nephew, had something to do with the butter's sale. The first line of figures includes some unexplained numbers, but note that "52 @ 34" means 52 pounds for 34 cents each pound. "Chs" - charges and "Coms" - commission are charged, as well as "Cartage." Compare the time between the day the tub was received and the day the receipt was written with the 1895 butter receipt shown below.

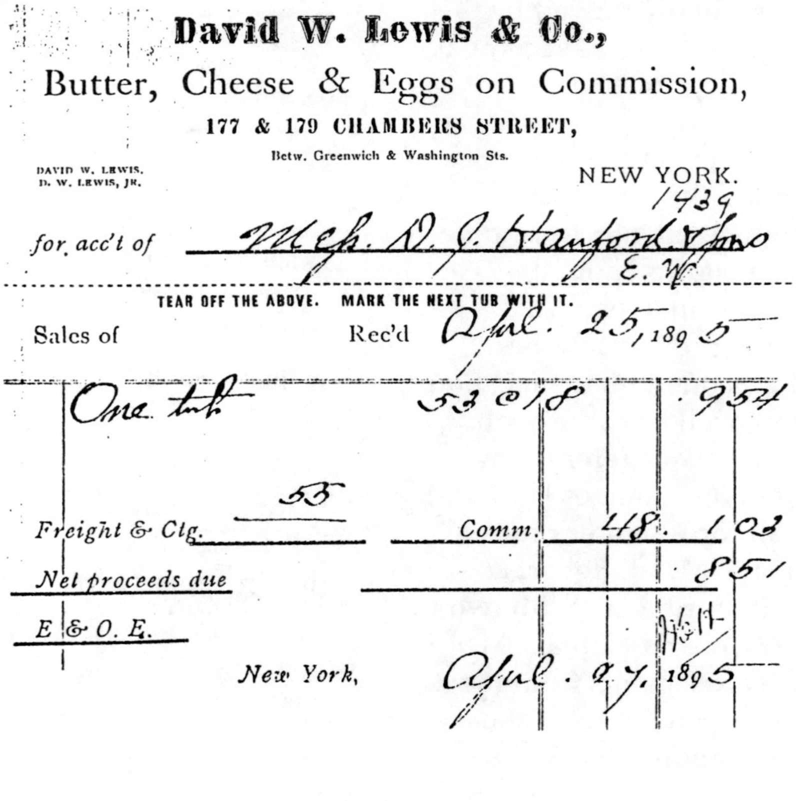

David W. Lewis & Co.,

Butter, Cheese & Eggs on Commission,

177 & 179 CHAMBERS STREET,

Betw. Greenwich & Washington Sts.

DAVID W. LEWIS.
D. W. LEWIS, JR.

NEW YORK.

1439

for acc't of Mess. D. J. Hanford & Sons
E. W.

TEAR OFF THE ABOVE. MARK THE NEXT TUB WITH IT.

Sales of Rec'd Apl. 25, 1895

One tub	53 @ 18	9 54
Freight & Ctg. 55	*Comm.* 48	1 03
Net proceeds due		8 51
E & O. E.		

New York, Apl. 27, 1895

This receipt is easier to read, though it includes the same information given in the above receipt. Here one tub is recorded as "53 @ 18" - 53 pounds sold at 18 cents a pound. As above, freight and a commission are deducted from the bill, but no cartage as the railroad was only three miles from East Meredith.

Tea Cows & Winter Butter - A Couple New Definitions

As winter came and cows dried up, butter would have become scarce if it weren't for "tea cows" and "winter butter."

Elizabeth Hanford chose to make winter butter. She often mentioned it in her diaries in the Fall of the year.

Saturday, October 13, 1883

Quite Warm. South Wind Heavy Shower at night Baby fell off the table on the floor. hurt him quite bad. James Leal very sick. Commence seventh tub for winter Butter

Elizabeth Hanford packed this winter butter in tubs and stoneware jars. She stored it in the cellar so it could be used throughout the winter months at home. Based on what she wrote in her diaries, Elizabeth kept anywhere from fifty to one hundred pounds of butter to be used at home as "winter butter." She also made winter butter for her local customers Mrs. Evans of East Meredith and P.R. Young of Oneonta.

If both butter and milk was required through the winter, another method had to be used. Milk would not keep in a cool cellar through the winter like butter could. At least one local resident in East Meredith remembers that her family kept what they called a "tea cow." This cow was bred later than the rest of the herd. This way, the tea cow gave birth late enough that her lactation period would make it possible to milk her throughout the winter. The cow's milk could provide cream for butter and milk for *tea* and other food items in the winter. The trade-off, though, was that the tea cow required better and more feed than the dry cows to help her produce milk for use.

More New Definitions

Here a few more terms related to bovines and dairy cows. In this urban age where fewer and fewer people work on farms or grow their own food, these words may not be familiar:

Neat is another word for bovines or cattle. It was often used in late 1800s census records together with the word cattle, thus *neat cattle*. In the past, the word "cattle" had referred to all domesticated animals considered as property. Thus, bovines, horses, goats, sheep and swine were all considered cattle of different sorts. The word "neat" was necessary to differentiate bovines from other types of cattle. The word "cattle" will hereafter mean bovines.

A ***Steer*** is a male bovine which has been castrated. In the United States, the word "steer" is often used when talking about any male cattle raised for beef.

Ox is another name for steer, in terms of being a castrated male bovine. The term "ox," or the plural "oxen," is most often used when referring to cattle used to pull wagons or farm equipment.

Female cattle only become ***Cows*** after they have their first calf.

A ***Heifer*** is a female bovine that has not had a calf.

A ***Milch Cow*** is a cow that can be used to give milk.

A ***Springer Cow*** is a cow that is ready to give birth, and thus, is nearly ready to milk.

A ***Stripper Cow*** is a cow at the end of her lactation period. This means that she will soon dry up and stop giving milk.

A ***Farrow Cow*** is a cow that has had a calf in the past, but when bred again, she failed to conceive another calf. Even though she had not given birth in the next lactation period, she may still give milk.

THE WORLD'S COLUMBIAN EXPOSITION & BUTTER

Did East Meredith's butter stay in the state of New York? For the most part, it did. At least once, however, East Meredith sent its butter further afield, and it was on view for nearly the whole world!

In 1893, Chicago, Illinois hosted the World's Columbian Exposition to celebrate the 400th anniversary of Columbus discovering North America. This World's Fair was one of the most popular world's fair to that date. Built near Chicago's waterfront, it had buildings to house exhibits on national and world industries, agriculture, women, art and many other topics as well as separate buildings for every state in the union. There were nearly 250,000 exhibits with over 70,000 exhibitors. Over 21 million people attended the fair between its opening on May 1 and its closing on October 30, 1893.

Part of these exhibits included a Dairy and Dairy Products for the State of New York exhibit. Dairy products from across the state of New York, including butter from Delaware County (which encompasses the East Meredith area), were put on display. World's Columbian Exposition records show that eleven East Meredith farmers sent their butter to Chicago:

J.P. Blakely
J.P. Brown
H. Clun
James Conner
George H. Conner
A. Every

WORLD'S COLUMBIAN EXPOSITION.
NEW YORK
DAIRY EXHIBIT.

L.R. Frisbee
C.O. Hanford
D.J. Hanford
S. Hetherington
H.H. Mitchell
Joseph I. Roberts
Charles Smith

Nearly 36% of all the World's Columbian Exposition exhibitors received commendation from the Exposition committee. These exhibitors received a bronze medal and a printed award. At least two East Meredith people received this award. Joseph I. Roberts' award is still in existence and is owned by Hanford Mills Museum. The second award went to C.O. Hanford, actually his wife Caroline. Caroline was Elizabeth Hanford's daughter-in-law, and she was the one who took over Elizabeth's "butter business." Presumably, she learned to make quality butter from Elizabeth. Though Caroline Hanford's award is not known to exist today, the Delaware County Dairyman newspaper mentioned it on February 8, 1895, perhaps when she received the actual certificate.

Friday - - February 8, 1895

Mrs. C.O. Hanford has a diploma awarded her for a package of butter, from the Columbian committee of the world's fair in Chicago.

Using Butter

Once Butter was made, packed and sold, what happened to it then? Of course, it was used!

Not only did Elizabeth Hanford use her diaries to record her daily activities, but she also occasionally used them to copy recipes that she liked. Her 1866 diary has the largest collection of her recipes, while other diaries only have a recipe here and there. They usually have less information than today's recipes. In the 1800s, most cooks already knew the basics of cooking and baking. They used written recipes as memory aids; to help them remember ingredient lists and proportions. Some of Elizabeth Hanford's original recipes are reproduced here using her own words and spellings, along with some suggestions for making them in today's modern kitchen. Notice there was no such thing as margarine or shortening in Elizabeth's cooking. Butter ruled the day.

Fruit Cake

2 lb. of butter
2 " Shugar
2 " Flour
12 Eggs beat the yolk with the Shugar the white to a froath 2 wine glass of rose water or brandy nutmeg 1 teaspoon Soda dissolved in warm water beet well togather then add 2 lb of Currants 2 of choped Raisins this will make 6 cakes Spice of all kinds prefered

Elizabeth's 1866 diary included three recipes for fruit cake. This one had the most complete instructions.

~ A wineglass measures about 1/4 cup. Rose water can often be purchased in health food stores, but be sure to find "food grade" rose water. Rose water was the flavoring of choice before vanilla became popular.

~ In the 1800s, soda, better known as baking soda was often dissolved in a small amount of water before it was added to the mixture.

~ In another of her fruit cake recipes, Elizabeth mentioned using the spices cinnamon, cloves, and nutmeg. The amounts must be adjusted by taste.

~ These cakes would probably bake at 350° F, perhaps for as much as an hour. Check them as needed.

This 1866 recipe has a light cornmeal flavor that is quite good. This cake can be eaten plain or frosted.

~ The whites should be beaten until they hold stiff peaks.

~ The batter will fit in an eight inch round pan, and should be baked at 350° F for about 30 to 40 minutes.

Corn Starch Cake

1/2 cup Corn Starch
1/2 " Butter
1/2 " Sweet Milk
1 " Sugar
1 cup Flour
Whites of three Eggs
1/2 teaspoon Cream tartar 1/2 Soda

Aple Cake

2 cups of sweet dried sour [apples] will do soak one night. Chop & boil an hour in 2 cup Molasis 2 eggs 1 cup butter 1 cup sour milk 1 cup Shugar 3 heaping of Flour 1 heaping tea spoon of Soda all kind of spices bake slow

Elizabeth's 1866 "Aple Cake" recipe appeared again in the West Kortright Church Cookbook of 1901. West Kortright is about five miles south of East Meredith. The two recipes are nearly identical except the older did not include raisins. To the left is the original, older recipe.

~ On today's stoves, where heat is constant, the apples and molasses may not have to cook an hour. Use your own judgement.
~ For spices, the newer 1901 recipe used 1 teaspoon of cloves and 1 teaspoon of cinnamon.
~ A slow oven means 250° to 300° F.

This is the only recipe that Elizabeth recorded in her 1866 diary for a dish used as a meal. It is a basic meat loaf recipe with a few twists.

~ In 1866, "minced beef," now known as hamburger, would have been very lean. In 1866, the Hanfords probably raised multi-purpose cattle - for working, milk and meat. That is why Elizabeth added butter to the recipe.
~ The size of soda crackers that Elizabeth used is unknown at this point. Sometimes crackers were in squares of 4, so 6 crackers may actually translate to 24 of today's crackers. Experiment.
~ Nutmeg may also have been selected to be larger in these modern times, so less of today's nutmeg may be required.
~ This recipe would probably bake as today's meat loafs - at 350° F.

Minced Beef

3 1/2 lbs of minced beef raw. 6 soda Crackers rolled fine 3 eggs beaten 1 1/2 tablespoon of Salt 1 teaspoon of peper 1 small nutmeg grated 4 tablespoon of cream Butter the size of an Egg Mix all thoughly in a loaf bake well in a dripping Pan 1 1/2 to 2 hours basting as other meats

Steamed Puding

1 cup molases
1 " Sweet Milk
1 " Raisins
1 egg
1 table spoon butter
1 teaspoon Soda
2 cups flour
put in a puding dish steam two hours

This comes from Elizabeth's 1901 diary. It is more explicit in its directions in the same way cookbooks were becoming more specific in the same time period Little explanation is needed. If you do not know how to steam pudding, check a cookbook like the Joy of Cooking. Any good general cookbook should be helpful.

A Little Bit About Cheese Making

Before Elizabeth Hanford and her family moved to East Meredith in 1866, they lived about 15 miles further east. They had a large farm and milked 30 cows, instead of the 10 they had later. They moved to East Meredith to start a store when Levi, Elizabeth's husband, started to have medical problems. As Levi recovered and their son grew older, the Hanfords went back into the farming business, but never as extensively as their first farm.

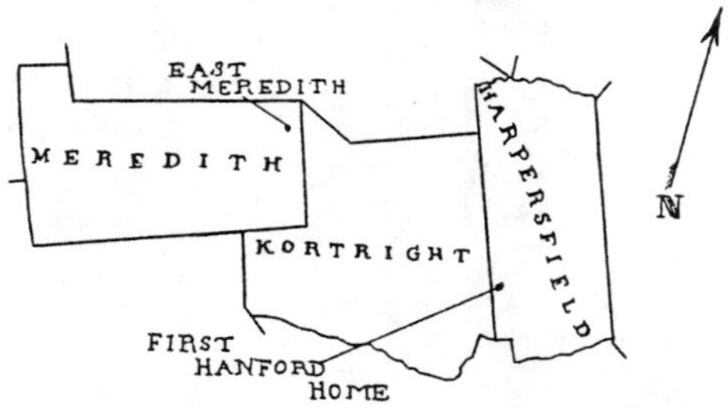

Elizabeth's early 1860s diaries show a different pattern of dairy production at their first and larger farm. In the spring, she packed butter in tubs as she did in East Meredith. As summer came on, though, she only packed a few firkins. Then she began to make cheeses through the summer and into the fall. She only stopped long enough to make winter butter for family use. Elizabeth made hard cheese, which requires more time and effort than soft cheese, but it keeps better. In 1863, she made at least 100 cheeses. She rarely recorded their size, but she did mention one cheese that weighed 80 pounds.

Mystery:

Why did East Meredith farmers make butter instead of cheese?

Was it because butter was easier and less time consuming to make?
OR
Was it because other areas had cheese factories that could produce cheese faster and at a lower cost?
OR
Did butter bring a higher price than cheese?

Was it one of these, a combination, or something altogether different?

So, what is involved in making hard cheese?

1. When enough milk is collected (an experienced cheese maker can get a pound of cheese from a gallon of milk), it must be heated to just below human body temperature. Next, rennet should be added. (Rennet is made from the stomach of a newborn calf, and necessary for making hard cheese.)
2. Let milk sit, up to one and a half hours, until the mixture jells into a solid jelly-like curd.
3. Cut curd gently into a checkerboard pattern. This releases a yellowish, clear whey.

4. Further cut the curd horizontally. If the whey starts to turn white, the curd is not being cut with enough care.
5. Push curd down, remove whey, heat it and pour back over the curd and stir.
6. Ladle mixture into cheese basket lined with cheese cloth to drain off whey. Whey made good pig feed.
7. Break up curd with hands.
8. Place curd in cheese hoop lined with wet cloth and place in cheese press. Press slowly at first, then harder. Leave in press for one to three days.

9. Remove cheese from press, rub with grease (usually butter or lard), and place on a shelf in a cool area to ripen. Turn the cheese daily and use a cloth to remove any mold that forms. Cheese will form a hard rind. Store until ripened or matured the desired length of time, at least six to eight weeks.

Chapter 4

THE END:

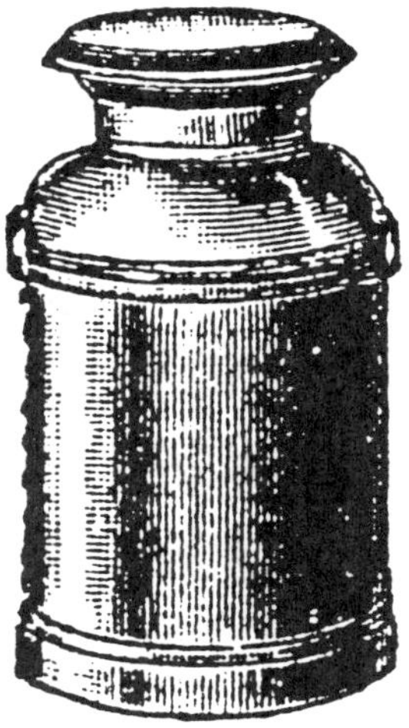

Changes in the Butter Business

After the turn-of-the-century, the demand for butter tubs and their covers began to decline. There were many factors that influenced this trend. The drive to earn more money and the growth in population, especially in urban centers, were overall reasons that had a part in all the factors that changed the Butter Business. Three main factors are listed below.

1. <u>The Railroad and Refrigerator Cars</u>
   As cities grew, their inhabitants required more and more food supplies. Railroads at first expanded their operations to meet growing demands and then developed new equipment to handle the increasing need. Food items, such as dairy products, began to travel farther and farther from their point of origin. As these innovations took place, the way foods were processed changed to meet the demands and the situations.

2. <u>Creameries</u>
   When the demand for dairy products began to grow and as farmers developed ways to increase output, higher amounts of food had to be processed in a uniform fashion. Dairy factories, known as creameries, could meet this challenge in processing dairy products much better than the local farmer. Not only could they process a larger volume of traditional products such as milk, cream and butter, but other less traditional dairy products. With special machinery, they could extract casein for early plastics and other items, produce dried milk powder and pasteurize large quantities of milk.

3. <u>New Agricultural Technology</u>
   In an effort to make farming less time consuming, and more productive and cost effective, agricultural technologies were always being improved. Better breeding, studies into feed nutrition and storage, milking machines and electricity are just four of the areas where advances were made that allowed agriculture to improve. They allowed farmers to increase their herds, their milking season, and, ultimately, to increase their production.

Though each of these factors were driven by a desire for more money and a growth in the population, but they were also interconnected. One was not possible without the others. There was no need to increase milk production if there wasn't someone to process it or transport it. There was no need for creameries if trains could not carry their products or if farmers couldn't produce the raw materials. The relationships are obvious and probably over simplified. No doubt, there are other factors that affected the changes in the Butter Business, but these were the most obvious to the local citizens of East Meredith.

THE RAILROAD & REFRIGERATOR CARS

As New York City and other Urban Centers expanded, it became increasingly difficult to obtain fresh dairy products. To solve the problem, railroads began to ship milk into the cities as early as the 1840s. The first cars were basically boxcars, sometimes fitted out with shelves to hold milk cans.

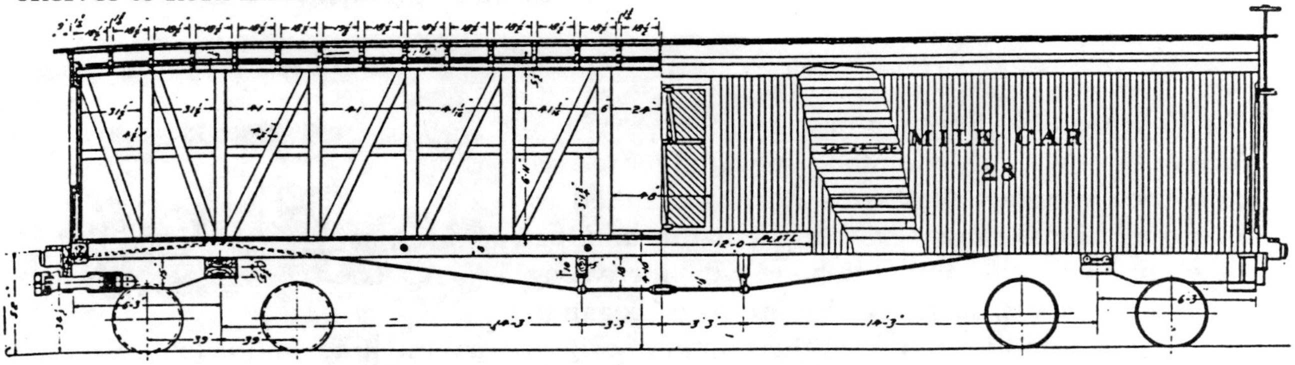

Since trains still had to travel only short distances, milk usually arrived in the city still fresh - except during the heat of summer. To solve this problem, milk was precooled and many milk trains ran at night to avoid the worst of summer's heat.

By the 1870s, as railroads had to travel further for milk, refrigerator cars were being used to transport milk. Surprisingly, refrigerator cars, or reefers as they were known, were in use by 1842 to transport seafood to parts of New England. A number of methods were used to cool the early refrigerator cars. They were usually insulated with bunkers or chambers built at the ends, in the roof, or in the middle to hold ice. As the train stopped at various stations, the bunkers were replenished with ice from railroad icehouses. As water melted, it left the car through drains under the car.

Although tank cars were being used to carry petroleum products also by the 1870s, milk was not carried in bulk until 1903. In that year, a tank was mounted in a refrigerator car. The experiment was considered a success, but the idea was not widely accepted until the 1920s.

So if refrigerated milk cars were available by the 1870s, why weren't East Meredith area farmers sending their milk out on the railroad in those years?

The answer is related to distances. The closest rail line to East Meredith in the 1870s was at a place called Cooperstown Junction on the Delaware and Hudson Railroad. To reach this station, an East Meredith area farmer would have to load his milk cans into his horse-drawn wagon and travel through hilly terrain over ten miles one way to the railroad.

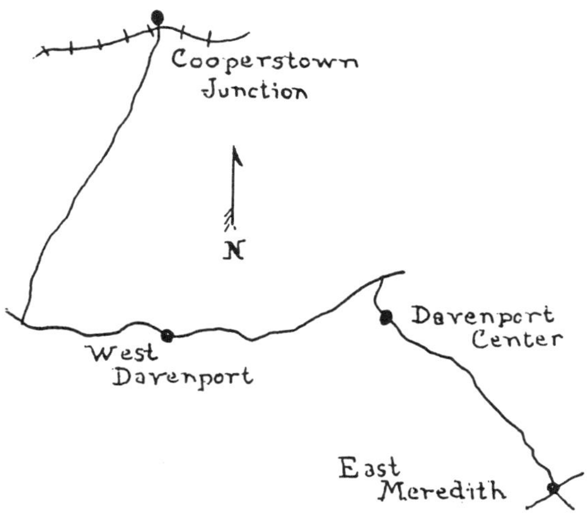

The 1870s wagon route from East Meredith to the Cooperstown Junction railroad station.

Not only could milk sour even before it reached a train, but the trip was at least three to four hours long one way. It was cost efficient to hire a driver or teamster to transport compact tubs of butter processed to withstand spoilage, but milk was bulkier and more susceptible to damage. In the 1890s, a rail line came within four miles of East Meredith, but even at that distance, milk transport was not convenient. It wasn't until the Ulster and Delaware Railroad reached East Meredith in 1900, that transportation of milk was realistically possible for local farmers.

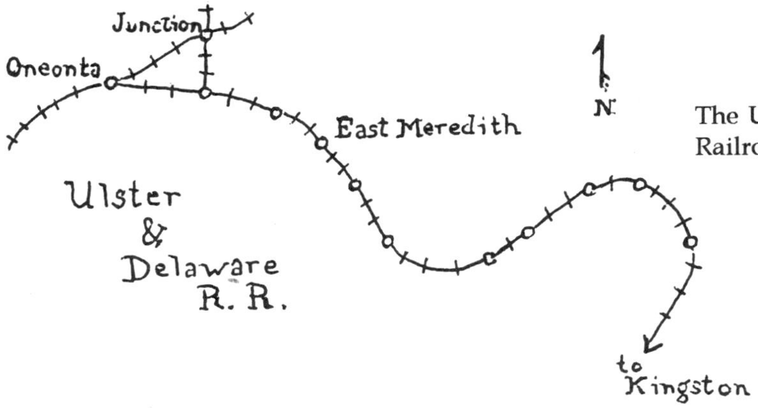

The Ulster and Delaware Railroad in 1900.

CREAMERIES - The Milk Processing Plant

The building where milk was processed into cream and butter on the farm was often called a creamery. The farm creamery should not be confused with the larger, factory-style creameries that local coopers perceived as the reason why they were being driven out of business. This letter, sent to the Hanfords in 1903, is perhaps the best testament to how the creameries affected the producers of butter tubs and butter tub covers:

> *211 Nield St.*
> *West Chester Pa.*
> *Mar. 7, 1903*
>
> *Mr. D.J. Hanford & Son.*
>
> *Dear Sirs: your letter of the 25th inst has reached me My address is as above and not 24 East Gay St.*
> *I never expected to get in such a hole, but the creameries coming in just knocked my business flat. I will tell you just how it is. My shop and tools stand there idel, but they are mortgaged for more than they are worth. I havent any sick relation to help me out of a hole and havent any thing to send you at present.*
> *I wrote you some time ago stating that I expected to pay the whole amount and that is still my intention now. By moving down here I have bettered my prospects and will be able to pay much sooner than if I had stayed in Durham. Of course I had to start in at a low salary just enough to live on at the price of fuel the past Winter, now I have an advance in position and as it is piece work will turn me in _?_ much more just as soon as I get more used to it. I am poor but love it and will pay you just as soon as possible.*
> *I appreciate your kindness in trusting me in the past and will prove to you later that you will not loose by it.*
>
> *Yours very sincerely,*
> *Walter R.T. Pratt.*

At least one other Hanford customer mentioned that his payments problems were caused by a creamery in the area, although he called it a milk station (see Chapter 3, George W. Sides - At the Beginning and the End). As the railroads and creameries spread, more farmers shipped milk instead of butter, and the butter tub trade dropped off.

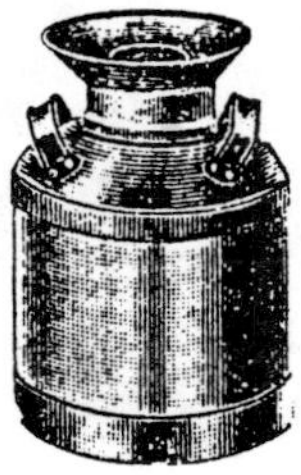

At the turn-of-the-century, the small village of East Meredith actually had two creameries built within its limits! Both were located along the Ulster and Delaware Railroad. The creamery on the east side of the village was known as the Empire State Creamery. Construction began in 1899 as the railroad was being completed, and it was ready for business when the first train came through in 1900. The Empire Creamery bottled local milk to ship. Around 1902, a second creamery was built on the northwest side of East Meredith. A number of farmers had been dissatisfied with the way the Empire Creamery did business, so they formed the Farmers' Co-Op and built the Co-Op Creamery. This creamery made and shipped butter, casein and pot cheese to New York City. Each creamery had delivery wagons that would travel to client farms within the area to pick up the milk each day. Later, motor trucks were used. The creamery processed the milk and paid the farmer for the milk that he supplied, less the cost of the delivery wagon.

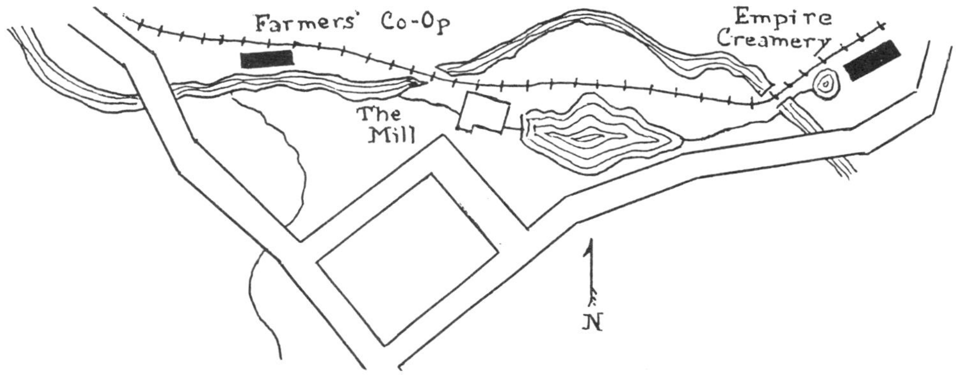

While creameries may have put butter tub manufacturing coopers out of business, this did not happen to the Hanfords' mill business. Horace Hanford, perhaps consciously or subconsciously perceived the trends, and began to manufacture and sell milk bottle crates and milk bottle draining racks even before the railroad arrived in East Meredith. His first few customers were creameries further away on other rail lines, but eventually it became a major part of the Hanford Mill woodworking business.

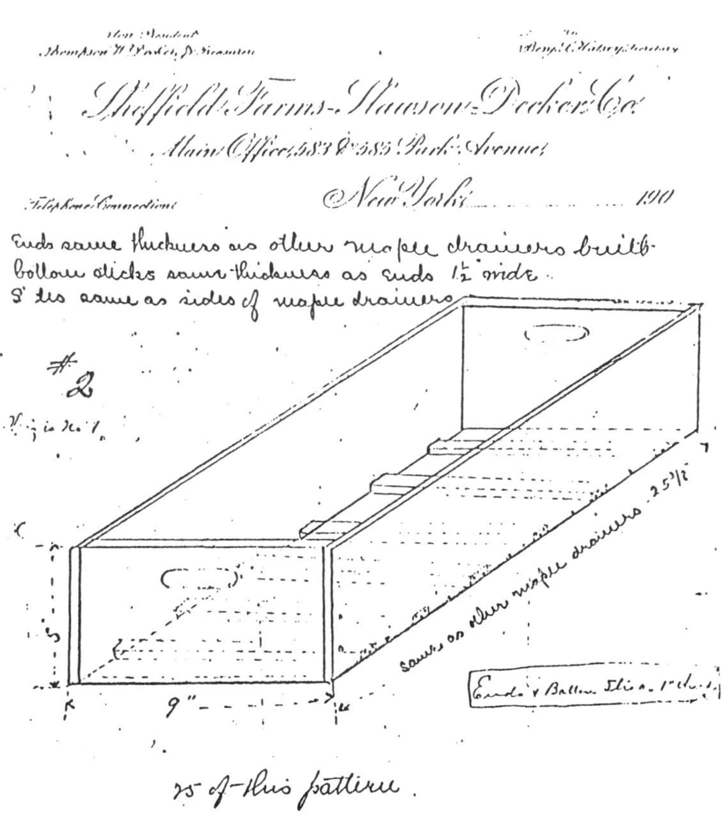

NEW AGRICULTURAL TECHNOLOGY

The growth in the production of milk was made possible by any number of new agricultural technological innovations. Here are just a few examples:

Nutrition and Silos

The major reason cows were dried off in the winter was because of nutritional needs. A cow producing milk needs more nutrients than a dry cow. One of the best feeds available for milk production is green, fresh grass, but that cannot be obtained in the winter. For most of the 1800s and before, cows lived on dried hay through the winter, supplemented perhaps with root crops, such as beets or turnips. The silo changed all that.

A silo is a building in which green fodder or animal feed is stored. In the East Meredith area the fodder is usually corn stalks. The fodder or silage, as it is called, ferments while maintaining a high nutritional value. Forms of the silo were used in Europe on and off for hundreds of years though it never gained complete popularity. The idea of the silo wasn't introduced to North America until 1875. There were many problems with early silos, so in 1882 there were only 92 farmers that owned them. By the 1890s, though, silos were making their way into the East Meredith area as this item from the local Delaware County Dairyman newspaper shows:

> Friday - - February 1, 1895
>
> Leonard Gibson, who has quite a large winter dairy, says the silo pays. He makes a large quantity of butter and gets a good price for it, even in these dull times.

With the silo, a farmer could breed different groups of cows in his herd at various times so he would be milking some portion of his cattle all through the year. Large quantities of milk, cream and butter became available throughout the year. As more and more farmers put in silos and milked cows throughout the year, they could no longer command high prices for summer and fall butter sold in the winter. Butter making was no longer as profitable as it had been, but there was a ready market for the increasing supply of milk.

Better Breeding

As farmers turned from subsistence farming to the production of food for commercial sale, there was incentive to improve the quality of their diary herd in a formal way. Farmers began to keep records of the milk and butter output of each cow, and which bulls and cows bred the most productive offspring. This drive toward more thoughtful breeding helped to increase the output of dairy cattle. This helped farmers meet the growing demand for more dairy products. Today, better breeding is carried to an extreme made possible only by modern technology. With artificial insemination, farmers no longer have only the local bulls to choose from, they can breed their prize cow to almost any top quality bull in the nation.

Milking Machines and Gasoline Engines

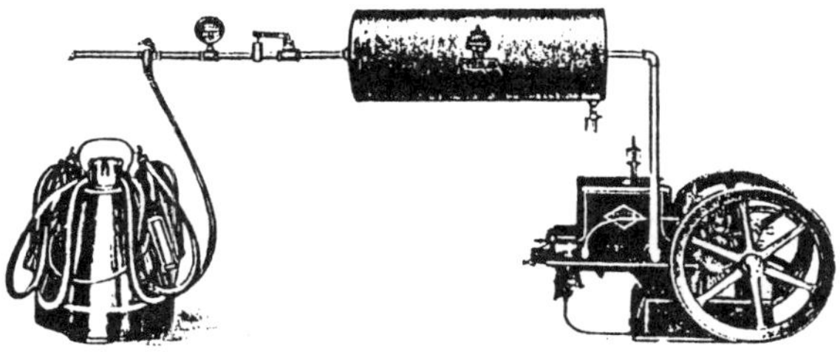

As mentioned in the previous section (see Chapter 3, Tools & Machinery - Miscellaneous Equipment and Tools & Machinery - Power Sources), farmers were experimenting with milking machines since at least the mid-1800s. With the advent of the affordable gasoline powered engine, the milking machine came into its own. More cows could now be milked faster with less human help. Farmers began to increase their dairy herd size without having to increase the number of farm hands they needed. According to one milking machine company, a good average for hand milking was eight cows in an hour, while their milking machine could milk twenty four cows in an hour. It didn't take farmers long to see the advantages.

Electricity

This last innovation didn't come to most East Meredith area farms until the 1940s. Some farms were able to get electricity earlier if they were on the edge of East Meredith and able to feed off the line that went through the village in 1926, while others bought home electric plants powered by gasoline engines. With electricity, milking machines could be run without an engine. Electric milk coolers and, later, bulk milk tanks could be installed, allowing the farmer to produce and store more milk.

OLEOMARGARINE - The Fake Butter

While many factors caused the decline of butter production in the East Meredith area, it was still produced there and in many other areas of the United States. It wasn't truly pushed to second place until wars and the health conscious encouraged the use of a substitute, a product known as oleomargarine, often called margarine today.

Oleomargarine was invented as a butter substitute by a French chemist, Hippolyte Mege-Mouries, in the late 1860s. Although ratios and specific ingredients vary, Mege-Mouries' recipe is little changed today. Margarine is usually made of fats and oils (animal and/or vegetable) mixed with milk, salt, and often yellow coloring. This oleomargarine was rapidly accepted in Europe, but when Mege-Mouries patented it in the United States in 1873 he ran up against heavy opposition. People in the dairy industry rightfully perceived oleomargarine as a threat to their butter sales. Dairy farmers lobbied congress, causing the federal government to pass the Oleomargarine Law in 1886. This law heavily taxed oleomargarine manufacturers, wholesale dealers, retail dealers and imported oleomargarine. Individual states also passed similar laws. Packages were required to be clearly marked so unsuspecting customers did not think they were buying butter. Since uncolored oleo was an unappetizing white color looking much like lard, it was often colored yellow to look like butter. Laws often required oleomargarine manufacturers to leave their product uncolored so buyers would not mistake it for butter. Manufacturers got around the problem by supplying customers with a small packet of yellow coloring that could be whipped into the oleo after they returned home with their purchase.

W., R. & Co.'s Improved Butter Color.
No. 7R2155 Large, $1.00 size, per bottle........80c
Medium, 50-cent size, per bottle.................38c
Small, 25-cent size, per bottle....................18c

The 1886 law and similar state laws curtailed the manufacture and use of oleomargarine. It wasn't until the World Wars, especially World War II, forced rationing of items that margarine saw any great number of sales.

A NATION AT WAR MUST BE A NATION OF SOLDIERS

EACH and every one of us must fight shoulder to shoulder.

Each doing that part in this great war which will bring a victory that will make wars forever a thing of the past.

To a few is given the privilege of fighting at the front where the glory and sacrifice are greatest.

With the rest of us lies the duty to see that our own men and the soldiers of our allies are supplied with the necessities that will end the war.

The United States Food Administration have asked the women, who are the soldiers of the kitchen, to save food.

One of the most important foods is wheat, which means bread.

The big bread supply comes from the bakers. They have been licensed and only the amounts of sugar, milk, shortening and wheat flour specified may be used, so as to economize on these important foods.

This will not help if the women, who bake at home, do not economize.

There are many delicious, wholesome breads that may be made with the unnecessary amounts of sugar and fat eliminated, or by using the many excellent vegetable oils that are on the market, in place of lard or butter.

OATMEAL MUFFINS

- 1 cake FLEISCHMANN'S Yeast
- ¼ cup lukewarm water
- 2 tablespoonfuls honey or corn syrup
- 1 teaspoonful salt
- 1 cup hot skim milk
- 1 cup rolled oats
- ½ cup whole wheat flour
- ½ cup sifted white flour
- 2 tablespoonfuls vegetable oil or butter substitute

BOIL oats and oil in milk one minute. Let stand until lukewarm. Dissolve yeast and sweetening in lukewarm water, add shortening and combine the two mixtures. Add flour and salt and beat well. The batter should be thick enough to drop heavily from the spoon. Cover and let rise until light, about one hour, in a moderately warm place. Fill well-greased muffin pans two-thirds full. Let rise about forty minutes, bake twenty-five minutes in a moderately hot oven.

From a World War I era Fleischmann's yeast pamphlet.

With butter rationed, consumers were forced to use more oleomargarine. Butter consumption went back up after the war, but by the 1950s, two things happened to change the trend again. The federal government repealed the 1886 Oleomargarine Law in that decade, and people started to become more health conscious. Vegetable based margarines were considered healthier than animal based butter. Butter producers did not give up easily, though. The last tax on margarine was repealed in Wisconsin only twenty two years ago, in 1967! Of course, there are still many people who still think baked goods are tasteless unless they are made with butter, and oleomargarine producers still try to make their product look and taste like butter.

AFTERWORD

I hope that you have enjoyed this brief tour through the Butter Business as it related to Hanford Mills and the area around East Meredith, New York. As it was not meant to be an all encompassing tome of butter history, I'm sure many points and topics were missed. I hope, however, that I have at least made it entertaining. Primary documents provided must of the information for this book, but there were other sources as well. The following books were consulted while putting this book together, and you may find them interesting. They deal with many topics closely or loosely related to butter.

~ Brockman, C. Frank. Important Forest Trees of the Eastern United States. (Washington, DC: United States Forest Service) 1968.

~ Courtney, W.S., revised by George E. Waring, Jr. The Farmers' and Mechanics' Manual. (New York: E.B. Treat & Co.) 1869.

~ Grafton, Carol Belanger, ed. Trades and Occupations: A Pictorial Archive from Early Sources. (New York: Dover Publications, Inc.) 1990.

~ Hedrick, Ulysses Prentiss. A History of Agriculture in the State of New York. (New York: Hill and Wang) 1966 edition.

~ Hornung, Clarence P., ed. Handbook of Early Advertising Art. (New York: Dover Publications, Inc.) 1956.

~ Johnson, Paul C. Farm Animals in the Making of America. (Des Moines, IA: Wallace Homestead Book Co.) 1975.

~ Kilby, Kenneth. The Cooper and His Trade. (Fresno, CA: Linden Publishing Co., Inc.) 1989.

~ Milener, Eugene. Oneonta: The Development of a Railroad Town. (Deposit, NY: Courier Printing Corp.) 1983.

~ Mitchell, Elma Hetherington. History of East Meredith, Volumes 1 through 3. (Laurens, NY: Village Printer) 1973, 1976 and 1986.

~ Patridge, Michael. Farm Tools Through the Ages. (Boston, MA: New York Graphic Society) 1973.

~ Proulx, E. Annie & Lew Nichols. The Complete Dairy Foods Cookbook. (Emmaus, PA: Rodale Press) 1982.

~ Salaman, R.A. Dictionary of Woodworking Tools c. 1700-1970 and Tools of Allied Trades. (Newtown, CT: Tauton Press) 1975.

~ Van Loon, Dirk. The Family Cow. (Charlotte, VT: Garden Way Publishing) 1981.

~ White, Jr., John H. The American Railroad Freight Car: From the Wood-Car Era to the Coming of Steel. (Baltimore, MD: Johns Hopkins University Press) 1993.

~ Wing, Henry H. Milk and Its Products. (New York: Macmillan Co.) 1908.

~ The 1901 West Kortright Church Cookbook. (East Meredith, NY: The West Kortright Centre) reprint, 1976.

~ The 1902 Edition of the Sears, Roebuck Catalogue. (New York: Bounty Books) 1969.

~ Asher and Adams' Pictorial Album of American Industry, 1876. (New York: Rutledge Books) reprint, 1976.

~ Biographical Review of Otsego County, New York. (Boston, MA: Biographical Review Publishing Co.) 1893.

~ Report of the Board of General Managers of the Exhibit of the State of New York at the World's Columbian Exposition. Transmitted to the legislature, April 18, 1894.